Corporate Power in Global Agrifood Governance

Food, Health, and the Environment
Series Editor: Robert Gottlieb, Henry R. Luce Professor of Urban and Environmental Policy, Occidental College

Keith Douglass Warner, *Agroecology in Action: Extending Alternative Agriculture through Social Networks*

Christopher M. Bacon, V. Ernesto Méndez, Stephen R. Gliessman, David Goodman, and Jonathan A. Fox, eds., *Confronting the Coffee Crisis: Fair Trade, Sustainable Livelihoods, and Ecosystems in Mexico and Central America*

Thomas A. Lyson, G. W. Stevenson, and Rick Welsh, eds., *Food and the Mid-Level Farm: Renewing an Agriculture of the Middle*

Jennifer Clapp and Doris Fuchs, eds., *Corporate Power in Global Agrifood Governance*

Corporate Power in Global Agrifood Governance

edited by Jennifer Clapp and Doris Fuchs

The MIT Press
Cambridge, Massachusetts
London, England

This book was set in Sabon by SNP Best-set Typesetter Ltd., Hong Kong. Printed and bound in the United States of America on recycled paper.

Library of Congress Cataloging-in-Publication Data

Corporate power in global agrifood governance / edited by Jennifer Clapp and Doris Fuchs.
p. cm.—(Food, health, and the environment)
Includes bibliographical references and index.
ISBN 978-0-262-01275-1 (hardcover : alk. paper)—ISBN 978-0-262-51237-4 (pbk. : alk. paper)
1. Farm produce—marketing. 2. Agricultural industries. 3. International business enterprises. 4. Globalization. I. Clapp, Jennifer, 1963– II. Fuchs, Doris A.
HD9000.5.c67 2009
382'.41—dc22

2008042146

10 9 8 7 6 5 4 3 2 1

Contents

Series Foreword

I am pleased to present the fourth book in the Food, Health, and the Environment series. This series explores the global and local dimensions of food systems and examines issues of access, justice, and environmental and community well-being. It includes books that focus on the way food is grown, processed, manufactured, distributed, sold, and consumed. Among the matters addressed are what foods are available to communities and individuals, how those foods are obtained, and what health and environmental factors are embedded in food-system choices and outcomes. The series focuses not only on food security and well-being but also on regional, state, national, and international policy decisions and economic and cultural forces. Food, Health, and the Environment books provide a window into the public debates, theoretical considerations, and multidisciplinary perspectives that have made food systems and their connections to health and the environment important subjects of study.

Robert Gottlieb, Occidental College
Series editor

Acknowledgments

This book originated with a workshop held in Waterloo, Ontario, in December 2006 on the theme of corporate power in global food governance. We would like to thank the Social Sciences and Humanities Research Council of Canada, the Centre for International Governance Innovation, and the Faculty of Environmental Studies at the University of Waterloo for financial support for that workshop. We are grateful to Clay Morgan and Sandra Minkkinen at the MIT Press for shepherding the project through the publication process, and to Robert Gottlieb and three anonymous reviewers for providing helpful comments. We would also like to thank Alex Bota, Matthew Bunch, Kim Burnett, Linda Swanston, and Candace Wormsbecker for research and editorial assistance. Finally, we would like to express our gratitude to the contributors to this book, as well as to our families and friends, for their patience and support throughout the publication process.

Contributors

Maarten Arentsen is Professor of Public Administration and Public Policy and Managing Director of the Centre for Clean Technology and Environmental Policy at the University of Twente, the Netherlands. His research focuses on energy policy, energy market reform, and (green) energy innovation, with a special focus on technological and institutional change. Among his recent publications are articles in *Energy Policy, Policy Studies Journal*, and *Energy and Environment.*

Jennifer Clapp is CIGI Chair in International Governance and Professor of Environment and Resource Studies at the University of Waterloo. Her current research projects focus on the global political economy of genetically modified food, the politics of food aid reform, transnational corporations and global food governance, and WTO agricultural trade negotiations. Among her recent books is *Paths to a Green World: The Political Economy of the Global Environment* (coauthored with Peter Dauvergne, MIT Press, 2005). She has also published articles in *Global Governance, Third World Quarterly, Journal of Environment and Development,* and *Ecological Economics*. She is coeditor of the journal *Global Environmental Politics.*

Robert Falkner is senior lecturer in international relations at the London School of Economics and associate fellow of the Energy, Environment and Development Programme at Chatham House. His research interests are in international political economy, with special emphasis on global environmental politics, multinational corporations, risk regulation and global governance. His most recent books are *Business Power and Conflict in International Environmental Politics* (Palgrave Macmillan, 2008) and *The International Politics of Genetically Modified Food: Diplomacy, Trade and Law* (Basingstoke: Palgrave Macmillan, 2007).

Doris Fuchs is Professor of International Relations and Development at the University of Muenster in Germany. She received her PhD in Politics and Economics in 1997 from the Claremont Graduate University and has since taught at the University of Michigan, Louisiana State University, the University of Munich, the University of Stuttgart, as well as the Leipzig Graduate School of Management. Her primary areas of research are private governance, sustainable development, food politics and policy, and corporate structural and discursive power. Among her publications are *Business Power in Global Governance* and *An Institutional Basis for Environmental Stewardship*, as well as articles in peer-reviewed journals such as *Millennium*, *Global Environmental Politics*, *International Interactions*, *Journal on Consumer Policy*, and *Energy Policy*.

Agni Kalfagianni is assistant professor at the University of Stuttgart, Chair of International Relations and European Integration. She has worked on the role of private and public actors in fostering sustainability and transparency in the food chain at the European and national (Dutch) levels. She has also published articles on corporate social responsibility strategies in European food and agricultural governance. Her main research interests include environmental policies and politics, sustainable development, environmental ethics, and democratic governance.

Peter Newell is Professor of Development Studies at the University of East Anglia and James Martin Fellow at the Oxford University Centre for the Environment. His most recent books include *The Business of Global Environmental Governance* (with David Levy, MIT Press, 2005) and *Rights, Resources and the Politics of Accountability* (with Joanna Wheeler, Zed, 2006). Among his other publications are articles in *Review of International Studies*, *Environmental Politics*, *Global Environmental Politics*, *Review of International Political Economy*, and *Journal of International Development*.

Steffanie Scott is Assistant Professor of Geography and Environmental Management at the University of Waterloo, Canada. She is involved in research on the restructuring of local food systems in Vietnam. Among her publications are articles in *Agriculture and Human Values*, *Regional Studies*, *Urban Geography*, *Women's Studies International Forum*, *International Development* Planning Review, and *Gender, Technology and Development*.

Susan K. Sell is Professor of Political Science and International Affairs and Director, Institute for International and Global Studies, at George Washington University. She has published three books on the politics of intellectual property —most recently *Intellectual Property: A Critical History* (coauthored with Christopher May, 2005). She has published articles in *International Organization*, *International Studies Quarterly*, *Global Governance*, and numerous law journals. Teaching and research areas include international political economy, international organization, nonstate actors, and international relations theory.

Elizabeth Smythe is Professor of Political Science at Concordia University College of Alberta in Edmonton, Alberta. Her research has focused on the negotiation of international trade and investment agreements and campaigns of resistance to them. Publications include "Democracy, Development and the Legitimacy Challenge: Assessing the Doha Development Round," in Donna Lee and Rorden Wilkinson, eds., *Endgame at the WTO: Reflections on the Doha Development Round* (Routledge, 2007), and "Legitimacy, Transparency and Information Technology: The World Trade Organization in an Era of Contentious Politics," *Global Governance* 12 (2006) (with Peter J. Smith).

Peter Vandergeest is Associate Professor in the Department of Sociology and Centre for Asian Research at York University, Toronto. His research has been organized around various dimensions of agrarian and environmental change in in Southeast Asia and includes studies of both forestry and agriculture. His current research takes up the history of alternative agriculture in Thailand and controversies around the privatization of environmental regulation in the global aquaculture industry.

Marc Williams is Professor of International Relations at the University of New South Wales, Sydney. His current research focuses on transnational civil society, trade politics, sustainable consumption, and the politics of genetically modified food. His most recent book (as coauthor) is *Global Political Economy: Evolution and Dynamics* 2nd edition (Palgrave, 2007). Among his other publications are articles in *Geopolitics, Global Environmental Politics, Environmental Politics*, and *Third World Quarterly*.

Mary Young completed her PhD dissertation at York University on globalization and agricultural restructuring in Indonesia. Her research interests include transnational agrofood production, political economy, and ecological change, particularly in Taiwan and Indonesia.

List of Acronyms

AAFC	Agriculture and Agri-Food Canada
AAN	Alternative Agriculture Network
AATF	African Agricultural Technology Foundation
ABIA	American Bio-Industry Alliance
ACT	Organic Agriculture Certification (Thailand)
ADDA	Agriculture Development Denmark (Asia)
ADM	Archer Daniels Midland
AP	Associated Press
APRESID	Asociación de Productores en Siembra Directa
ArPOV	Asociación Argentina de Protección de las Obtenciones Vegetales
ASA	American Soybean Association
ASA	Asociación de Semilleros Argentinos
AU	African Union
BIO	Biotech Industry Organization
BIOS	Biological Innovation for Open Society
BIOTECanada	Biotech Industry Organization Canada
BRC	British Retail Consortium
BSE	Bovine spongiform encephalopathy
CABC	Canadian Biotechnology Advisory Committee
CAC	Consumers Association of Canada
CAS	Cámara Argentina de Semilleros
CASAFE	Camara de Sanidad Agropecuaria y Fertilizantes
CBD	Convention on Biodiversity
CCFL	Codex Committee on Food Labelling
CEO	Chief executive officer
CFIA	Canadian Food Inspection Agency

CGIAR	Consultative Group on International Agricultural Research
CI	Consumers International
CIMMYT	International Maize and Wheat Improvement Center
CONABIA	Comisión Nacional Acesora de Biotecnología Agropecuaria
COP	Conference of the Parties
CSR	Corporate social responsibility
DANIDA	Danish International Development Agency
DFID	Department for International Development (UK)
DNMA	Dirección Nacional de Mercados Agroalimentarios
EC	European Commission
ETI	Ethical Trading Initiative
EU	European Union
Eurep	Euro-retailers Produce Working Group
FAB	Foro Argentino de Biotecnología
FAC	Food Aid Convention
FAO	Food and Agriculture Organization
FBT	Food, beverage, and tobacco
FCD	Fédération des entreprises du Commerce et de la Distribution
FDA	Food and Drug Administration
FDI	Foreign direct investment
FSC	Forest Stewardship Council
FTA	Free trade agreement
FWAG	Farming and Wildlife Advisory Group
GAO	Government Accountability Office
GAP	Good Agricultural Practices
GATT	General Agreement on Tariffs and Trade
GFSI	Global Food Safety Initiative
GlobalGAP	Global Partnerships for Good Agricultural Practice
GM	Genetically modified
GMA	Grocery Manufacturers Association
GMOs	Genetically modified organisms
GMP	Good Manufacturing Practices
GPL	General public license
GRAIN	Genetic Resources Action International

GTZ	German International Development Agency
GURTs	Genetic use restriction technologies
HACCP	Hazard Analysis of Critical Control Points
IAASTD	International Assessment of Agricultural Knowledge, Science and Technology for Development
IATP	Institute for Agriculture and Trade Policy
ICONE	Institute for the Study of Trade and International Negotiations
ICS	Internal Control System
ICTSD	International Centre for Trade and Sustainable Development
IFOAM	International Federation of Agricultural Movements
IFPRI	Food Policy Research Institute
IFS	International Food Standards
IGC	Intergovernmental Committee on Intellectual Property, Genetic Resources, Traditional Knowledge and Folklore
IGC	International Grains Council
ILSI	International Life Sciences Institute
IMO	Institute for Marketecology
INASE	Instituto Nacional de Semillas
INGOs	International nongovernmental organizations
IP	Intellectual property
IPM	Integrated pest management
IPR	Intellectual property rights
IRRI	International Rice Research Institute
ISAAA	International Service for the Acquisition of Agro-Biotech Applications
ISLI	International Life Sciences Institute
ISO	International Organization for Standardization
ITF	International Task Force on Harmonization and Equivalence in Organic Agriculture
IUPGR	International Undertaking on Plant Genetic Resources
KARI	Kenya Agricultural Research Institute
M4P	Making Markets Work Better for the Poor
mha	Million hectares

NAFTA	North American Free Trade Agreement
NAWG	National Association of Wheat Growers
NEPAD	New Partnership for Africa's Development
NGO	Nongovernmental organization
NIEO	New International Economic Order
OECD	Organization for Economic Cooperation and Development
PBRs	Plant Breeders' Rights
PCT	Patent Cooperation Treaty
PhRMA	Pharmaceutical Research and Manufacturers Association
PIPRA	Public Intellectual Property Resource for Agriculture
PR	Public relations
PTO	Patent and Trademark Office (US)
PVP	Plant Variety Protection
R&D	Research and development
rDNA	Recombinant DNA
RR	Roundup Ready
RSC	Royal Society of Canada
SAGPyA	Secretaría de Agricultura, Ganadería, Pesca y Alimentos
SENASA	Servicio Nacional de Sanidad y Calidad Agroalimentaria
SPLT	Substantive Patent Law Treaty
SPS	Sanitary and Phytosanitary Measures
TBT	Technical Barriers to Trade
TNCs	Transnational corporations
TRIPS	Trade Related Intellectual Property Rights
UK	United Kingdom
UNCTAD	United Nations Conference on Trade and Development
UPOV	Union for the Protection of New Varieties of Plants
US	United States
USAID	United States Agency for International Development
USDA	United States Department of Agriculture
U.S. PIRG	United States Public Interest Research Group

USTR	United States Trade Representative
USWA	United States Wheat Associates
WETEC	Wheat Export Trade Education Committee
WFP	World Food Program
WHO	World Health Organization
WIPO	World Intellectual Property Organization
WTO	World Trade Organization

1

Agrifood Corporations, Global Governance, and Sustainability: A Framework for Analysis

Jennifer Clapp and Doris Fuchs

Fundamental changes have taken place in food systems around the world over the past century. We now have a globally integrated food system that affects all regions of the world. The recent volatility in food prices has illustrated the global nature of this food system, highlighting the ways developments in one part of the world can have multiple and wide-ranging impacts. Transnational corporations (TNCs) have been central actors in the development of this global food system. They dominate the production and international trade in food and agricultural items, and are also key players in the processing, distribution, and retail sectors. Indeed, it is unlikely that the current global food system would exist as it does today without the participation of TNCs. Many of these firms operate in numerous countries and at more than one level along the global food chain. As food systems around the world become increasingly affected by the corporate-dominated global food system, we must pause to consider the consequences of this fundamental change in the provision of our food.

The international governance of the food system is geared toward providing some degree of regulation to put in place safeguards from potential negative socioeconomic and ecological consequences of a globalized food system. In many ways these rules govern activities of agrifood corporations, because these actors are pivotal agents in the globalization of the food system. But at the same time, these corporations play a key role in the establishment of the very rules that seek to govern their activities. This includes influence over state-based and intergovernmental mechanisms of governance, as well as private forms of governance. This situation raises important theoretical and policy questions about the impact of corporate influence on the sustainability—environmental as well as economic and social—of the global food system and the rules that govern it.

This book examines several key questions about the role of corporate actors in the global food system. First, we ask what role corporations are playing in the formation of the norms, rules, and institutions that govern the global food system. In answering this question, we examine the various facets of power that corporations exercise in an attempt to influence rules by which they themselves must ultimately play, including both state-based and private forms of governance. Second, we ask about the wider implications of corporate power in global food governance for the sustainability of the global food system as well as for societal debates over sustainability in the global food system. In particular, we examine the ways corporations use and influence the definition of the concept of sustainability in their exercise of power in global food governance in relation to other actors and other interpretations of the concept.

This book sits at the intersection of two emerging literatures on TNCs. Recent years have seen a growing body of scholarship on transnational corporations as actors in global governance (Fuchs 2007a; Cutler, Haufler, and Porter 1999; Cutler 1999; Sklair 2002). Grounded primarily in political science and international political economy, this work examines the political role of corporate and business actors in the establishment and implementation of norms, rules, and institutions governing international political and economic interactions. Power, authority, and legitimacy are central themes to this work. At the same time, a literature has emerged on the role of TNCs in the global agrifood system (see, for example, Glover and Newell 2004; McMichael 2005; Lang and Heasman 2004; Magdoff, Foster, and Buttel 2000; Heffernan 1998; Bonanno et al. 1994). Grounded primarily in sociology and political ecology, this literature examines the implications of growing corporate activity in the food system. The impacts of corporate concentration and market domination on society and the environment have been key themes in this literature.

Both of these literatures have been important in advancing our understanding of TNC activity in the global sphere. Yet while both focus on TNCs as actors in the global political economy, there has been little cross-referencing between these two literatures. As such there has been little academic work that specifically examines the *political* role that corporations play in efforts to *govern* the global *food* system. The literature on corporate actors in global governance has not yet examined the food and agriculture industry in any significant depth. The literature on TNCs in the food sector has focused on market power, with an implicit assumption that economic dominance translates into political clout. Market power is certainly important, but it does not capture all of the facets of power that corporations can exercise in their bid to shape the rules of the game. Given the current instability in the global food system and the central role TNCs play in it, there is an urgent need for much more systematic and comprehensive analysis of the political role that corporate interests play in global food governance.

The framework we develop here identifies different channels through which corporations influence global food and agriculture governance and examines the implications of that influence. By focusing on the intersection of TNCs and global food governance, the book aims to build on the existing literature on TNCs and the global food system, as well as on the conceptual literature on the role of corporations in global governance more broadly. A crucial contribution of this book is to unpack the complex relationship between the exercise of power and the use of the concept of sustainability in the governance of the global food system.

Following a discussion of the rise of TNCs in the global food system more broadly, this introductory chapter presents a conceptual framework to help explain the role and implications of transnational corporate involvement in the global governance of food and agriculture. The case studies presented in this book have been selected to allow as comprehensive an understanding of corporate involvement in global agrifood governance as possible. They include investigations of corporate influence with respect to standard setting by retail corporations, international food safety standard setting, food aid policy, regulation of genetically modified organisms (GMOs), global biosafety governance, and trade-related intellectual property rights rules. With this theoretically founded and empirically comprehensive approach, this book seeks to make significant progress in our understanding of global agrifood governance and the role of corporations in it.

Corporate Actors and the Rise of a Global Food System

The international trade in certain food and agricultural items has characterized food systems for centuries. But the truly global scale of production, trade, and marketing of food and agricultural products that we have today really only developed in the past fifty years (e.g., Busch and Juska 1997; Magdoff, Foster, and Buttel 2000; McMichael 2000, 2005;

Weis 2007). National food economies are increasingly integrated into a global food system founded on ever-growing volumes and value of agrifood trade. Although the share of agricultural trade in world merchandise trade declined between 1980 and 1997 from 17 to 10 percent (Ingco and Nash 2004, 5), the markets for global food and agriculture products have grown markedly in absolute terms, and have become much more globally integrated. There was a significant increase in the value of global agricultural trade overall during this period as well. The FAO reports that international agricultural trade is expanding more rapidly than world agricultural output (FAO 2005, 12).

At the same time that international trade in agricultural products has grown, there have been shifts in its makeup both in terms of countries and products. Over the 1971–2001 period, the growth in food imports was most marked in developing countries, which saw a rise of 115 percent, compared to developed countries with a rise of 45 percent (FAO 2004, 16). Developing countries were net agricultural exporters in the 1960s, but now are net agricultural importers (FAO 2004, 14). Since the early 1980s, trade in processed food also has grown more quickly than trade in bulk commodities. Today, trade in processed food products accounts for some 66 percent of agricultural trade (FAO 2004, 26).

Parallel to the growth in agricultural trade has been a growing participation of TNCs in the food and agriculture sector (Heffernan 2000; Murphy 2006). Indeed, global trade in food and agricultural products takes place largely among TNCs (FAO 2003). In the mid-1970s, there was heightened concern over the increasingly global scope of the grain trade and the power of corporate players in that trade (Morgan 1979). Since that time, corporations have diversified into multiple facets of the food sector, including commodity trading, food processing and retailing, as well as seed and agricultural chemical production (FAO 2003). TNCs have been central players in the global integration of the food system. They have stretched their operations both vertically and horizontally, to the point that it no longer makes sense to speak of national food systems because the agrifood TNCs are so globally integrated in their operations (Heffernan 1998). As part of this globalization process, agriculture and food have become commodified through complex and global production chains dominated by TNCs, which demand durable products and thrive on distance, both social and physical, between the production and consumption of food (Friedmann 1993, 1994; Kneen 1993).

In the industrialized world—and increasingly in the less industrialized world as well—a growing proportion of the food people eat has an international and corporate dimension. At least one of the steps in the food chain– from production, trade, processing, and packaging to retailing—is typically overseen by a major food corporation (see Lang and Heasman 2004). Moreover, the liberalization of foreign direct investment rules in developing countries in recent decades has facilitated the rapid expansion of supermarkets in the global South, most of which are owned by major international retail corporations. This shift toward truly global supermarket retailing has meant more concentration in the procurement and distribution of foods in both rich and poor countries (FAO 2005, 21–22; Burch and Lawrence 2007; Konefal, Mascarenhas, and Hatanaka 2005).

Indeed, corporate involvement at all stages along the food production chain (inputs, production, commodity trade, processing, and retailing) has become much more concentrated in recent years (Heffernan 2000). According to the FAO (2003, 119), "a small number of companies now dominate each part of the food chain in OECD countries." High ratios of concentration are present in many sectors within countries, and this concentration is also a feature of the global system. Just four companies control nearly 40 percent of the global coffee trade, while three roasting companies control 45 percent of the global coffee market (FAO 2004, 30). A report by the Erosion, Technology and Concentration (ETC) Group documents the level of concentration in the various stages of the food and agriculture production chain:

- The top 10 seed companies control nearly 50 percent of the US $21 billion annual global commercial seed market and nearly all of the genetically engineered seed market.
- The top 10 pesticide companies control 84 percent of the US $30 billion annual global pesticide market.
- The top 10 food retailers control 24 percent of the estimated US $3.5 trillion global food market.
- The top 10 food and beverage processing companies control 24 percent of the estimated US $1.25 trillion global market for packaged foods. (ETC Group 2005)

Increasing corporate concentration in the global agrifood system has provoked many questions from both within and outside of academia about corporate accountability and responsibility in this sector (e.g., Murphy 2006; ActionAid International 2005; MacMillan 2005).

McMichael (2005, 280) argues that this increasingly globally oriented corporate control over the food system is resulting in a privatization of food security, whereby we see "an emerging *world* agriculture subordinated to capital." The push for agricultural trade liberalization under programs of structural adjustment in the global South as well as via the Agreement on Agriculture of the World Trade Organization (WTO) are seen by many to have pushed the global food system yet further in this direction (McMichael 2005, 292; Weis 2007).

The globalization of food and agricultural systems may have produced some benefits, such as increased varieties of foods available to consumers and new markets for producers. But critical thinkers and a number of NGOs have also raised concerns about the impacts of corporate control of "food globalization" on socioeconomic and environmental outcomes. In particular, there is a growing critique of the effects that corporate concentration in a globalized food system is having on food security, small-farmer livelihoods, environmental quality, food safety, and consumer sovereignty. These concerns have been driving forces behind international efforts to establish rules, norms, and institutions to govern the global food and agricultural systems. Global food governance rules and institutions supposedly seek to ensure that each of the steps in the food chain is carried out in a manner that mitigates risks and maximizes benefits of the internationalization of the food system (Phillips and Wolfe 2001). At the same time, however, corporate actors in the global agrifood system are increasingly playing a role in setting the very rules that govern their activities. This raises further concerns about the efficacy and legitimacy of these rules.

Private, corporate actors have taken on a significant role in the global food system, not only as economic actors responsible for much of the world's food production, processing, and retailing, but also as political actors in global mechanisms to govern the food system. But does the economic significance and growing concentration among agrifood TNCs translate into political power in global governance mechanisms? And what is the significance of corporate involvement in global food governance in terms of outcomes?

To answer these questions it is important to understand the multiple facets of power that corporations exercise in global governance more broadly, and to investigate whether these facets of power are visible in their activities with respect to global food governance. It is also important to examine the ways in which the power of corporate actors influences outcomes, particularly with respect to questions of sustainability (Fuchs 2007b). In the following sections, we outline the various facets of corporate power in global governance and begin to map out how they unfold with respect to governance of the global food system.

Conceptualizing Corporate Power in Global Food Governance: Toward a Framework

Recent years have spawned numerous studies on different aspects of the political role of business and corporate participation in global governance (Clapp 2005; Cutler 1999; Haufler 2002; Levy and Newell 2004). Supplementing earlier scholarship in international political economy on the structural power of TNCs (Gill and Law 1989), this recent work has looked at the rise of "private authority" in the global political realm and its wider implications (Cutler, Haufler, and Porter, 1999; Fuchs 2005a, 2007a; Grande and Pauly 2005; Falkner 2003).

Several main themes emerge from the literature on the political role of corporations. First, strategies pursued by international industry lobby groups in their attempts to frame the politics of specific issues represent an important object of inquiry (Clapp 2003; Rowlands 2000). Second, the ways in which the structural power of corporate actors in the broader global political economy influences governance outcomes need to be considered (Fuchs 2007a; Levy and Egan 1998; Newell and Paterson 1998). Third, the "legitimacy" of corporations as political actors and the "authority" they can exercise vis-à-vis the state and civil society—in particular via the development of private-industry-driven governance institutions and the discursive framing of policy issues- require attention (Fuchs 2005b; Falkner 2003). These themes provide a starting point for investigations into the role of corporations in global agriculture and food governance.

The understanding of corporate power employed in much of the literature on globalization, including that in the food sector, tends to equate corporate power with market share (e.g., MacMillan 2005). While the economic dimension of corporate influence is important, our approach is to go deeper to uncover the different political facets of corporate power and its sources. A multifaceted approach to corporate power reveals the many ways this power is being employed, which together constitute corporations' ability to influence the governance of the global food system. Specifically, these different facets of power enable corporations to have a say in what is on the agenda and what is not, and to shape the distribution of the costs and benefits of the resulting rules and regulations. Market power influences political power, but one cannot assume that it translates into political power in a one-to-one relationship. Accordingly, it is important to unpack corporate power and to look at its different political facets as well as to consider important additional sources besides market power such as access to information and the policy process, or the perceived political legitimacy of corporate actors. Following on earlier work with respect to the dimensions of power in global governance more broadly, we pay particular attention here to the instrumental, structural, and discursive dimensions of corporate power (see also Fuchs 2005a, 2007a; Levy and Egan 2000).

One of the more obvious forms of power held by corporate actors is the instrumental power they attempt to wield in policy processes via corporate lobbying or political campaign financing.[1] This type of power can be characterized as direct influence of one actor over another, and assumes the existence of a functional, unilinear causality; of individual voluntary action and instrumental causality (e.g., Dahl 1957). Specifically, scholars link this type of power to a change in the decision outcomes by one actor due to the influence exercised by another. Actor-specific resources, such as financial, organizational, or human resources, as well as access to decision makers, all feed into this type of power. In political science, this instrumental power has been at the center of analyses of pluralism and interest-group politics. Such instrumentalist approaches to power can also be found in traditional power theories in the field of international relations, where scholars focus on the use of power by states in pursuit of national interests.

While an instrumentalist perspective is important in explaining direct power of one actor over another, it fails to capture the power exercised via the imposition of limits on the range of choices given to actors. To more fully understand the way corporations might exert control over actors' choice sets, a structuralist perspective on power needs to be added to the framework. This "second face of power" stresses the importance of the input side of the political process and of the predetermination of the behavioral options of political decision makers by existing material structures that allocate direct and indirect decision-making power. This structuralist perspective on power takes into consideration the broader influence corporate actors have over setting agendas and making proposals as a product of their material position within states and the global economy more broadly. In doing so, it emphasizes the importance of examining the contexts that make alternatives more or less acceptable before the actual and observable bargaining starts (Bachrach and Baratz 1970). The structural power TNCs derive from the ability to punish and reward countries for their policy choices by relocating investments and jobs has been the focus of considerable attention by critical international political economy scholars (Cox 1987; Gill and Law 1989; Fagre and Wells 1982).

A further, more dynamic type of structural power has also emerged in recent years. As globalization has continued, material structures have increasingly put corporate actors in a position to make governance decisions themselves, either supplementing or in some cases replacing traditional actors such as states and global institutions. Economic and institutional structures, processes, and interdependencies have created a setting where corporate actors have control of pivotal networks and resources. This control has given them the capacity to adopt, implement, and enforce privately set rules that may then take on an obligatory quality and that also have distributional consequences for others (Fuchs 2005a, 2007a). Some scholars have pointed to the development of private and quasi-private regimes, such as the International Organization for Standardization's ISO 14000 standards, whereby industry players took a key role in not only setting the agenda, but also in delineating the rules (Clapp 1998). When privately developed regimes are adopted or encouraged by states as a form of "regulation," or given legitimacy by international organizations such as the World Trade Organization, they take on broader public significance, even though they were not developed in the public sphere. Firms' growing adherence to "corporate social responsibility" (CSR) and private certification standards, and the growing significance they play in the regulatory structures governing the global economy, illustrate the importance of understanding this aspect of corporate power. This type of power is structural in the sense that it affects the input side of the political process. In other words, it allows corporate actors to determine the focus and content of rules. Likewise, it is structural in that their position in global economic networks provides corporations with a strong influence on which private standards and labels are widely adopted.[2]

Even an analysis of both the instrumental and the structural facets of a corporate actor's power, however, does not fully explain the power that preexists decisions and nondecisions. A more discursive approach, which recognizes that power is also a function of norms and ideas and is reflected in discourse, communicative practices, and cultural values and institutions, is necessary as well (Lukes [1974] 2004). Scholars attach increasing importance to this discursive dimension of the political process. They point out that policy decisions are, to a growing extent, a function of discursive contests over the framing of policies and the assignment of problems into categories by linking them to specific fundamental norms and values (Hajer 1997; Kooiman 2002).

Discursive perspectives draw out two main insights on corporate power. First, power not only pursues interests, but also creates them. Power in this sense precedes the formation and articulation of interests in the political process because of its role in constituting and framing policies, actors, and broader societal norms and ideas. Corporate actors, in other words, often play an important role in framing certain issues and problems in public discourse. This discourse in turn can have an important yet indirect influence over the ways public debates are carried out and over the choices presented to society for addressing them. Corporate lobby groups, such as the World Business Council on Sustainable Development (later transformed into the Business Action for Sustainable Development), played a prominent role in issue framing around the Rio Earth Summit as well as at the Johannesburg World Summit on Sustainable Development. Its engagement in the global dialogue on globalization and the environment portrayed voluntary corporate measures as more user-friendly and efficient, not to mention more environmentally sound, than government regulation (Finger and Kilcoyne 1997; Clapp 2005).

A focus on discursive power allows for a richer analysis of the presence and exercise of power in the absence of observable conflicts of interests. Conceptualizing power in this way provides a basis for a more comprehensive explanation of how political systems prevent certain demands from becoming political issues. This "third face of power" considers attempts to socialize politicians and the public into accepting "truths" about desirable policies and political developments and pays attention to media and public relations efforts, among others (Lukes [1974] 2004).[3] While such attempts to influence the public perception of key issues may sometimes backfire, this discursive engagement can have profound impacts on public debate.

The second insight from discursive understandings of power is that legitimacy is intimately wrapped up in discourse itself (Fuchs 2005b). In other words, in order to effectively exercise discursive power in the political process, an actor requires political legitimacy. After all, discursive power is relational in that it relies on the willingness of message recipients to listen and to place a least some trust in the validity of the contents of the message. Political legitimacy, in turn, can derive from a variety of sources. Public actors generally have this legitimacy on the basis of electoral processes and the formal authority associated with political offices. Actors can also obtain political legitimacy from the trust the public places in their expertise and capacities as well as in their intentions. These latter sources of political legitimacy apply primarily to nonstate actors and are frequently discussed in the literature on private authority.[4]

An analysis of the role of corporations in global agrifood governance, then, needs to consider the instrumental, structural, and discursive facets of corporate power. Analyses should pay attention to the role played in the governance of the global food system by lobbying and associated financial payments and contributions, agenda-setting power derived from capital mobility, rule-setting activities reflected in self-regulation and public-private or private-private partnerships, and efforts to influence the public debate through the framing of policy issues and underlying societal norms.

Some caveats to our three-dimensional concept of power need to be mentioned. First, it is important to note that the clear differentiation between the different dimensions of power is a simplification for analytic purposes. Empirically, the dimensions of power are more difficult to separate. Moreover, they interact and can even enhance each other (Fuchs 2007a). Second, this three-dimensional concept of power is one of a number of concepts of power currently employed in the field of international relations. Susan Strange's notion of "structural power," for example, included various sources of power—namely, finance, production, security, and knowledge (Strange 1988). Her view of structural power, then, combined several elements of power that we discuss here, in particular market-based power (what we call structural power) as well as discursive power (as part of the knowledge structure). While they are indeed related, we feel that it is important to differentiate more explicitly between these types of power. Barnett and Duvall (2005) have suggested a four-dimensional concept of power, differentiating between compulsory, institutional, structural, and productive power. In our view, however, the lines between some of these dimensions of power are all too blurred, while other important aspects of power, captured particularly in the notion of discursive power, do not receive sufficient attention.

Thus, we believe that the three-dimensional approach to power with its differentiation between instrumental, structural and discursive power holds the most explanatory value.

In addition, potential relationships between corporate power and the power of other actors, such as states and nongovernmental organizations, need to be taken into account. While it often makes sense to employ an analytic simplification and investigate business and state power separately, they ultimately interact in complex and multifaceted ways, and are frequently interdependent (Falkner 2003). Thus, one has to ask whether American TNCs are particularly powerful actors in global governance due to the predominance of the United States among the state actors. Similarly, state and TNC actors operate closely on many regulatory fronts, as is seen in the agrifood sector, for example with regard to trade policy, food safety, and biotechnology (see, e.g., Newell and Glover 2004). Some have pointed to the "revolving door" between industry and the state in the agrifood sector, where high-level employees of the state leave government to take on industry positions, and vice versa, affecting the making of governmental policy and regulation (e.g., Murphy 2006; Newell 2003). Similarly, there are a number of instances where NGOs and industry are working together with respect to setting industry standards (Nadvi and Wältring 2002). In this context, the potential instrumentalization of the state and other actors by private actors needs to be considered.

By offering a differentiated analysis of the instrumental, structural, and discursive facets of this power in global food governance, this book aims to contribute to the literature on corporate power. Specifically, the various chapters all analyze at least two of these facets in the context of specific issues associated with the governance of the global food system. In addition, they explore the interaction of corporate power with the power of other actors in their specific areas of analysis and the ways this interaction may enhance or constrain corporate power.

Implications of Corporate Power for Sustainability in the Global Food System

It is widely agreed that the organization of food and agriculture systems must be a key component of policies to promote both socioeconomic and environmental sustainability (e.g., IAASTD 2008; World Bank 2007). There are sound reasons for this centrality of food and agriculture in discussions of sustainability. Over half of the world's population is engaged in agricultural production, making it a source of livelihood for a significant proportion of people on the planet. Agricultural production is also intimately tied to the land and thus has direct environmental consequences. Farmers are often inadequately recognized for their stewardship of almost one-third of the Earth's land (IAASTD 2008). Yet while there is agreement that food and agriculture policy are vital components of sustainable development policy, there is considerable debate over exactly how the global food and agriculture system should be organized in order to best promote socioeconomic and environmental goals. For example, there are widely differing viewpoints on the relative merits of industrial agriculture versus small-scale farming, genetically modified seeds versus organic production, and global versus local food systems (see, e.g., Weis 2007; Ansell and Vogel 2006; Falkner 2007; Lang and Heasman 2004).

The contested nature of the concept of sustainability has enabled it to be used strategically by different actors (Fuchs 2007b). As such, the institutions and arrangements for governing the global food system have become sites of debate over how best to organize food and agricultural systems to promote sustainability. The exercise of corporate power and influence is important to consider in this respect. The various facets of corporate power can play a role in setting the parameters and tone of debates about the sustainability implications of various models of organization for global food and agriculture. In the past decade in particular, corporate actors have become much more active in attempting to influence these debates as public concern has risen over food safety and the introduction of genetically modified (GM) foods (Falkner 2007; Ansell and Vogel 2006).

At various points along the food production chain, governments and multilateral institutions oversee the governance of various aspects of the global food system. These mechanisms include international agreements and other institutional arrangements. These global efforts have been important sites of debate over ecological and socioeconomic issues—in terms of the need to set rules to mitigate negative consequences of the global food system in the absence of rules, but also in terms of the impacts of the rules themselves.

At the agricultural-input end of the spectrum, corporations and environmental groups sparred over the extent to which the Cartagena Protocol on Biosafety should restrict trade in GMOs in order to protect the environment. At the same time, it has become clear that the agreement has not been able to prevent the unwanted spread of GMOs (e.g., Glover and Newell 2004; Clapp 2006a). There has also been active debate about whether stronger intellectual property rights rules under the Trade Related Intellectual Property Rights (TRIPS) Agreement provide actual benefits for developing countries (e.g., Sell 1999). And environmental groups have accused international agricultural assistance institutions of bending under corporate pressure to encourage the use of industrial agricultural practices associated with the Green Revolution—many of which have been associated with environmental problems such as soil erosion and loss of genetic diversity—throughout the developing world (e.g., Shiva 2000).

With respect to agricultural commodity trade, there has been extensive debate over whether and the extent to which rules as established in the WTO Agreement on Agriculture exacerbate inequities relating to agricultural trade and protectionism between rich and poor countries (e.g., Clapp 2006b; Rosset 2006). There has also been critique of corporate attempts to influence the current governance of international food aid (Zerbe 2004; Barrett and Maxwell 2005). On the food processing and retail fronts, the Codex Alimentarius has been accused of setting standards that facilitate trade and meet the needs of agrifood corporations more than they promote food safety per se (Sklair 2002; Suppan 2006).

Private industry and third-party-led initiatives in the global food and agriculture system are also sites of debate. Reporting efforts that reflect corporate social responsibility have been increasingly adopted by major firms in the food and agriculture sector, which argue that it demonstrates their commitment to sustainability. Some have raised doubts, however, about how much CSR initiatives contribute to development and environmental and social protection in this sector (e.g., Tallontire 2007; Clapp 2008). Private, industry-led standards, which sometimes are third-party certified or involve NGOs in standard setting, have also emerged to promote food safety and quality, as well as certification for organic and fair trade products. The large transnational grocery retail corporations, for example, are increasingly demanding that their suppliers meet quality and safety standards, and this requirement filters all the way back through processors, traders, and down to the level of production by farmers. But the proliferation of private certification schemes is seen by many to be pushing small farmers out of the market, particularly those operating in the developing world, in favor of large agribusiness and food processors (Hatanaka, Bain, and Busch 2005; McMichael and Friedmann 2007).

Concerns about the various institution-, industry-, and third-party-led governance efforts have raised important questions about the ability of these measures to effectively promote sustainability in all its dimensions. Some scholars see corporate concentration in the global food system as directly leading to rules that reinforce a global and corporate-led agrifood system (McMichael 2005). Many regard a corporate-controlled global food system as antithetical to sustainability because it contributes to an industrial agricultural model that relies on heavy chemical use and agricultural biotechnology (IAASTD 2008). Corporate concentration is further seen to foster inequality as well as a reliance on international trade for food security, which many believe contributes to heightened vulnerability (e.g., Norberg-Hodge, Merrifield, and Gorelick 2002). Such critiques have spawned movements such as those promoting "food sovereignty," and "local food movements," which seek to retreat from the global, corporate-led food and agriculture system, to a more local and ecologically sustainable food system. Such movements reject current governance mechanisms, such as the WTO Agriculture Agreement, and also seek to promote "GMO-free zones" in the face of perceived weaknesses of the Biosafety Protocol (e.g., Rosset 2006; McMichael 2005).

In this context of debate over the sustainability dimensions of food and agriculture, it is important that the roles TNCs are playing in that debate are further unpacked and examined. The framework we set out in the previous section helps to better understand that role and in turn to better understand the origins of the corporate positions in the broader debate. The framing of the discourse around agricultural technologies and sustainability itself on the part of agrifood corporations, the structural power corporations hold in the market, the influence of lobby power held by a handful of agrifood corporations, and self-set voluntary rules and certification standards, all play into the framing of sustainability in this broader debate as well as sustainability outcomes as such.

The Plan of This Book

The case studies presented in this book document the channels through which corporations today shape global agrifood governance, and the implications of this role for sustainability and the broader debate over agricultural sustainability. The various chapters investigate a large range of political activities by corporate actors, the linkages between a corporation's economic and political power, the interaction between corporate power and state power, as well as the existing constraints on corporate power in the global governance of the agrifood system.

The book is divided into two parts. Part I provides insights into the nature of corporate power in various areas of global food governance and its implications. Part II then focuses on one particular area of global food governance: the governance of genetically modified organisms, which have received particular attention in recent years both in analyses of the global food system in general and in studies of the role of corporations in the global food system. The case studies in both sections contribute unique insights on corporate power in global food governance and its implications for debates over socioeconomic and environmental characteristics of the global food system.

In chapter 2, Fuchs, Kalfagianni, and Arentsen explore the increasing power of retail corporations and its implications for sustainability. Starting from the recognition that a new type of retail corporation has arrived on the scene and produced a shift in power between the actors involved in global agrifood governance, Fuchs and colleagues delineate the structural and discursive power of retail corporations today. Specifically, they examine retail corporations' rule-setting power as expressed in private standards and the exercise of discursive power reflected in the strategic use of quality and sustainability frames. They then investigate the implications of those forms of power for sustainability, considering issues of food safety, as well as environmental and social sustainability. It is with respect to the latter, especially, that Fuchs, Kalfagianni, and Arentsen highlight the danger of dramatic costs to societies, especially in developing countries. In consequence, they discuss opportunities for the regulation of retail power in the food sector as well as for reducing the negative effects of private retail governance on sustainability.

Scott, Vandergeest, and Young continue the discussion of food marketing in chapter 3 with their examination of certification standards for "green" foods in Southeast Asia. They argue that because transnational corporations are key players in setting and enforcing standards that are required for local firms to export organic foods from this region to international markets, these players have significant structural power. Corporations also have discursive power due to the way they frame organics as a key marketing point for Western markets. While corporations are key players in the politics of certification standards for these "green" foods in Southeast Asia, this chapter highlights their complex relationship with other actors, including the state, NGOs, farmers' associations, aid agencies, and social movements in this fast-growing international market for organic foods.

In chapter 4, Smythe examines the role and extent to which corporate power affects emerging norms of transparency at both the global and national level and thereby raises questions about democratic accountability and the rights and capacity of food consumers to have choices. She specifically addresses corporate power in the rule-making process of the Codex Alimentarius. Consumers in many countries are increasingly distant and detached from the sites of crop production and processing. As food production has become globalized, it has fallen under this complex array of governing institutions and private corporate networks, raising questions about the nature and legitimacy of the rules that are established and the interests they serve. As Smythe shows, the voices of consumers and small-scale local producers are increasingly at risk of becoming marginalized in this governance system. One reaction to globalization and global governance has been a growing demand for increased transparency in the processes of governance. However, this demand is as yet unfulfilled.

In chapter 5, Clapp analyzes the role of corporate power in international food aid policy, with a focus on the United States. She argues that corporate actors exercise instrumental power at the national level through their lobbying activities in the US Congress regarding international food aid policy. Moreover, she shows how they exercise structural power, which results from their economic and national security significance in the United States. In addition, she documents that corporations exercise discursive power in food aid governance by engaging in public debate over questions of hunger and food security and its link to food aid. Finally, Clapp delineates the complex and changing relationship between corporations and other nonstate actors with a stake in food aid policy.

Part II starts out with Williams's analysis in chapter 6 of the promotion of GMOs by corporate actors. Specifically, he examines discursive strategies pursued by individual corporations as well as industry associations in attempts to frame GM food, highlighting how the discursive power of corporate actors is being used to construct a positive normative consensus around GMOs as an alternative to the discourse of risk employed by GMO opponents. In this context, the chapter examines the promotion of "environmental sustainability" and "food security" as key normative frames, using official statements and publications by leading biotechnology firms. Here, Williams shows that the power to determine what counts as knowledge, which actors disseminate it, and the terms on which it is communicated has an enormous impact on public policy-making. Finally, the chapter examines interventions designed to promote the spread of GM crops and the consumption of GM food in Africa, in particular.

In chapter 7, Sell examines the role global biotechnology firms have played in efforts to raise regulatory standards for intellectual property protection worldwide and its implications. She points out that the complexity of the regulatory environment has given corporations important advantages in achieving their objectives, because vertical and horizontal forum shifting between multilateral, regional, and bilateral venues and across international organizations has made it difficult for developing countries to keep abreast of intellectual property policymaking and has made them vulnerable to power plays from global corporations and their supportive trade ministries such as the Office of the United States Trade Representative. Sell delineates the impact of this process on developing-country agriculture, highlighting the instrumental, structural, and discursive dimensions of the contemporary policymaking environment. In addition, she explores the potential outcomes of the ongoing contest between corporate actors on the one side and NGOs and developing-country governments on the other for the setting of IPR-related norms in the global economy and the emerging alternatives for developing-country agriculture.

In chapter 8, Falkner investigates the contestation of corporate power in the governance of GMOs. He shows that the growth of agricultural biotechnology has not been straightforward, pointing out resistance by consumers, food producers, retailers, farmers, and regulators as well as the creation of biosafety regulations at the national and international level. On this basis, Falkner questions the actual extent of corporate power in global food governance as well as the potential of social and political checks on international business and technology. In pursuit of these objectives, Falkner examines recent cases of contestation in agribiotechnology that have shaped the seemingly unstoppable march of biotechnology. He finds that business conflict plays a key role in limiting the power of the corporate sector and opens up crucial political spaces for other actors.

In chapter 9, Newell inquires into the role of corporate power in the governance of GMOs in Argentina. This country is a leading exporter of GM foods, a key ally of other biotech superpowers like the United States and Canada, and a potential gateway to the rest of the Latin American continent for biotech products. He shows that the story of agricultural biotechnology in Argentina brings to the fore a powerful combination of the politics of poverty, the power of transnational corporations, and the political economy of food and agriculture. Following a brief overview of developments in GMO politics and policy in the country, the chapter provides insights into the role of key private actors in this sector, and describes their corporate and policy strategies and forms of association and mobilization. Newell then discusses different dimensions of corporate power at play, before closing with some tentative conclusions about the implications of these forms of corporate power for alternative ways of organizing the governance of food and agriculture in a context in which state economic strategy is so entirely dependent on private agricultural interests.

Chapter 10, the book's conclusion, draws together the common and differing insights from the various case studies and discusses the complex implications of corporate power in global agrifood governance for sustainability. In addition, it explores the role of the structural context of this power, the relevance of other actors, and the role of issues such as knowledge. The chapter highlights interesting strategies employed by corporations as well as factors influencing the likelihood of success of the corporate exercise of power. The chapter then points out strategies for limiting negative effects of corporate power on the sustainability of global agrifood governance. It closes with a discussion of the implications of the insights on global food governance gained for assessing and improving its democratic legitimacy, addressing questions of participation, transparency, and accountability.

Indeed, based on an investigation of the role of corporate power in global agrifood governance and its implications for sustainability, issues of democratic legitimacy—that is, questions of participation, transparency, and accountability—remain as the central issues in global food governance. While public actors are not necessarily "good" nor private actors necessarily "bad" forces in governance, public actors generally are selected by more participatory processes and have to be accountable to more interests and criteria than private ones. Accordingly, the book's documentation of the large role corporate actors play in global agrifood governance today raises the questions of how to select the new "governors" and how to make them accountable to those governed. By providing a first set of answers to the questions of the corporate power in global agrifood governance and its consequences, then, the book and the case studies gathered in it serve to guide future research on the topic of agrifood corporations and global food governance toward pivotal issues for the future of humankind: the issues of the sustainability and democratic legitimacy of the global food system and global food governance.

Acknowledgments

The authors would like to thank the Social Sciences and Humanities Research Council of Canada, the Centre for International Governance Innovation, and the Faculty of Environmental Studies at the University of Waterloo for funding the workshop that started off this project. We would also like to thank three anonymous reviewers for helpful comments.

Notes

1. Note, however, that in critical state theories, the term *instrumental power* is used to refer to specific mechanisms of business control of state policy. Examples include the "revolving doors" and social networks that facilitate the lobbying and campaign finance frequently associated with claims of a substantial degree of business control of state policy levers (Miliband 1969).

2. In other words, corporations can punish and reward NGO-created standards and labels—for instance, for their choice of criteria—just as the traditional notion of structural power identified the ability of TNCs to punish and reward governments for their policy choices.

3. That media and public relations efforts would be among the key foci of such an analysis makes clear that an actor's material resources matter in the exercise of discursive power.

4. It is important to note that the underlying notion here is one of output legitimacy rather than input legitimacy—that is, a legitimacy focusing on results rather than notions of participatory democratic norms and procedures (Scharpf 1998).

References

ActionAid International. 2005. *Power Hungry: Six Reasons to Regulate Global Food Corporations*. Johannesburg: ActionAid. http://www.actionaid.org.uk/_content/documents/power_hungry.pdf.

Ansell, Christopher, and David Vogel, eds. 2006. *What's the Beef? The Contested Governance of European Food Safety*. Cambridge, MA: MIT Press.

Bachrach, Peter, and Morten Baratz. 1970. *Power and Poverty: Theory and Practice*. New York: Oxford University Press.

Barnett, Michael, and Raymond Duvall, eds. 2005. *Power in Global Governance*. Cambridge: Cambridge University Press.

Barrett, Christopher, and Daniel Maxwell. 2005. *Food Aid After Fifty Years: Recasting Its Role*. London: Routledge.

Bonanno, Alessandro, Lawrence Busch, William Friedland, Lourdes Gouveia, and Enzo Mingioni, eds. 1994. *From Columbus to ConAgra: The Globalization of Agriculture and Food*. Lawrence: University Press of Kansas.

Burch, David, and Geoffrey Lawrence, eds. 2007. *Supermakets and Agri-food Supply Chains: Transformations in the Production and Consumption of Foods*. Cheltenham: Edward Elgar.

Busch, Lawrence, and Arunas Juska. 1997. Beyond Political Economy: Actor Networks and the Globalization of Agriculture. *Review of International Political Economy* 4 (4): 688–708.

Clapp, Jennifer. 1998. The Privatization of Global Environmental Governance: ISO 14000 and the Developing World. *Global Governance* 4 (3): 295–316.

Clapp, Jennifer. 2003. Transnational Corporate Interests and Global Environmental Governance: Negotiating Rules for Agricultural Biotechnology and Chemicals. *Environmental Politics* 12 (4): 1–23.

Clapp, Jennifer. 2005. Global Environmental Governance for Corporate Responsibility and Accountability. *Global Environmental Politics* 5 (3): 23–34.

Clapp, Jennifer. 2006a. Unplanned Exposure to Genetically Modified Organisms: Divergent Responses in the Global South. *Journal of Environment and Development* 15 (1): 3–21.

Clapp, Jennifer. 2006b. WTO Agriculture Negotiations: Implications for the Global South. *Third World Quarterly* 27 (4): 563–577.

Clapp, Jennifer. 2008. Illegal GMO Releases and Corporate Responsibility: Questioning the Effectiveness of Voluntary Measures. *Ecological Economics*. 66 (2–3): 348–358.

Cox, Robert. 1987. *Production, Power, and World Order*. New York: Columbia University Press.

Cutler, Claire. 1999. Locating "Authority" in the Global Political Economy. *International Studies Quarterly* 43 (1): 59–81.

Cutler, Claire, Virginia Haufler, and Tony Porter. 1999. *Private Authority and International Affairs*. Albany: SUNY Press.

Dahl, Robert. 1957. The Concept of Power. *Behavioural Science* 2 (July): 201–215.

ETC Group. 2005. Oligopoly, Inc. 2005. *Communiqué* 91, November/December. www.etcgroup.org/upload/publication/pdf_file/42.

Fagre, Nathan, and Louis Wells. 1982. Bargaining Power of Multinationals and Host Governments. *Journal of International Business Studies* 13 (2): 9–23.

Falkner, Robert. 2003. Private Environmental Governance and International Relations: Exploring the Links. *Global Environmental Politics* 3 (2): 72–87.

Falkner, Robert. 2007. *Business Power and Conflict in International Environmental Politics*. London: Palgrave Macmillan.

Finger, Matthias, and James Kilcoyne. 1997. Why Transnational Corporations Are Organizing to "Save the Global Environment." *The Ecologist* 27 (4): 138–142.

Food and Agriculture Organization. 2003. *Trade Reforms and Food Security: Conceptualizing the Linkages*. Rome: FAO. ftp://ftp.fao.org/docrep/fao/005/y4671e/y4671e00.pdf

Food and Agriculture Organization. 2004. *State of Agricultural Commodity Markets 2004*. Rome: FAO. ftp://ftp.fao.org/docrep/fao/007/y5419e/y5419e00.pdf.

Food and Agriculture Organization. 2005. *State of Food and Agriculture 2005: Agricultural Trade and Poverty: Can Trade Work for the Poor?* Rome: FAO. ftp://ftp.fao.org/docrep/fao/008/a0050e/a0050e_full.pdf.

Friedmann, Harriet. 1993. After Midas's Feast: Alternative Food Regimes for the Future. In Patricia Allen, ed., *Food for the Future: Conditions and Contradictions of Sustainability*, 213–233. New York: Wiley.

Friedmann, Harriet. 1994. Distance and Durability: Shaky Foundations of the World Food Economy. In Philip McMichael, ed., *The Global Restructuring of Agri-Food Systems*, 258–276. Ithaca, NY: Cornell University Press.

Fuchs, Doris. 2005a. Commanding Heights? The Strength and Fragility of Business Power in Global Politics. *Millennium* 33 (3): 771–803.

Fuchs, Doris. 2005b. Governance by Discourse. Paper presented at the Annual Meeting of the International Studies Association, Honolulu, March.

Fuchs, Doris. 2007a. *Business Power in Global Governance*. Boulder: Lynne Rienner.

Fuchs, Doris, with Jörg Vogelmann. 2007b. Business Power in Shaping the Sustainability of Development. In Jean-Christophe Graz and Andreas Noelke, eds., *Transnational Private Governance and Its Limits*, 71–83. London: Routledge.

Gill, Stephen, and David Law. 1989. Global Hegemony and the Structural Power of Capital. *International Studies Quarterly* 33 (4): 475–499.

Glover, Dominic, and Peter Newell. 2004. Business and Biotechnology: The Regulation of GM Crops and the Politics of Influence. In Kees Jansen and Sietze Vellema, eds. *Agribusiness and Society: Corporate Responses to Environmentalism, Market Opportunities and Public Regulation*, 200–231. London: Zed.

Grande, Edgar, and Louis Pauly, eds., 2005. *Complex Sovereignty: Reconstituting Political Authority in the 21st Century*. Toronto: University of Toronto Press.

Hajer, Maarten. 1997. *The Politics of Environmental Discourse: Ecological Modernization and the Policy Process*. Oxford: Clarendon Press.

Hatanaka, Maki, Carmen Bain, and Lawrence Busch. 2005. Third Party Certification in the Global Agrifood System. *Food Policy* 30 (3): 354–369.

Haufler, Virginia. 2002. Globalization and Industry Self-Regulation. In Miles Kahler and David Lake, eds., *Governance in a Global Economy*, 226–252. Princeton, NJ: Princeton University Press.

Heffernan, William. 1998. Agriculture and Monopoly Capital. *Monthly Review* 50 (3): 46–59.

Heffernan, William. 2000. Concentration of Ownership and Control in Agriculture. In Fred Magdoff, John Bellamy Foster, and Frederick Buttel, eds., *Hungry for Profit: The Agribusiness Threat to Farmers, Food, and the Environment*, 61–75. New York: Monthly Review Press.

Ingco, Merlinda, and John Nash. 2004. What's at Stake? Developing Country Interests in the Doha Development Round. In Merlinda Ingco and John Nash, eds., *Agriculture and the WTO*. Washington, DC: World Bank.

International Assessment of Agricultural Knowledge, Science and Technology for Development (IAASTD). 2008. *Executive Summary of the Synthesis Report of the International Assessment of Agricultural Knowledge, Science and Technology for Development*. Washington, D.C.: IAASTD. http://www.agassessment.org/docs/SR_Exec_Sum_210408_Final.pdf.

Kneen, Brewster. 1993. *Land to Mouth*. Toronto: NC Press.

Konefal, Jason, Michael Mascarenhas, and Maki Hatanaka. 2005. Governance in the Global Agri-Food System: Backlighting the Role of Transnational Supermarket Chains. *Agriculture and Human Values* 22 (3): 291–302.

Kooiman, Jan. 2002. Governance: A Socio-Political Perspective. In Jürgen Grote and Bernard Gbikpi, eds., *Participatory Governance. Political and Societal Implications*, 71–96. Opladen: Leske + Budrich.

Lang, Tim, and Michael Heasman. 2004. *Food Wars: The Global Battle for Mouths, Minds and Markets*. London: Earthscan.

Levy, David, and Daniel Egan. 1998. Capital Contests: National and Transnational Channels of Corporate Influence on the Climate Change Negotiations. *Politics and Society* 26 (3): 337–361.

Levy, David, and Daniel Egan. 2000. Corporate Political Action in the Global Polity. In Richard Higgott, Geoffrey Underhill, and Andreas Bieler, eds., *Non-State Actors and Authority in the Global System*, 138–153. London: Routledge.

Levy, David, and Peter Newell, eds. 2004. The Business of Global Environmental Governance. Cambridge, MA: MIT Press.

Lukes, Steven. [1974] 2004. *Power, a Radical View*. London: Palgrave Macmillan.

MacMillan, Tom. 2005. *Power in the Food System: Understanding Trends and Improving Accountability*. Background Paper. Food Ethics Council. http://www.foodethicscouncil.org/files/foodgovreport.pdf.

Magdoff, Fred, John Bellamy Foster, and Frederick Buttel, eds. 2000. *Hungry for Profit: The Agribusiness Threat to Farmers, Food, and the Environment*. New York: Monthly Review Press.

McMichael, Philip. 2000. The Power of Food. *Agriculture and Human Values* 17 (1): 21–33.

McMichael, Philip. 2005. Global Development and the Corporate Food Regime. In Frederick Buttel and Philip McMichael, eds., *New Directions in the Sociology of Global Development*, 265–299. Amsterdam: Elsevier.

McMichael, Philip, and Harriet Friedmann. 2007. Situating the Retail Revolution. In David Burch and Geoffrey Lawrence, eds., *Supermakets and Agri-food Supply Chains: Transformations in the Production and Consumption of Foods*, 154–172. Cheltenham: Edward Elgar.

Milliband, Ralf. 1969. *The State in Capitalist Society*. New York: Basic Books.

Morgan, Dan. 1979. *Merchants of Grain*. New York: Viking.

Murphy, Sophia. 2006. *Concentrated Market Power and Agricultural Trade*. Ecofair Trade Dialogue, Discussion Paper No. 1, August. http://www.tradeobservatory.org/library.cfm?refid=89014.

Nadvi, Khalid, and Frank Wältring. 2002. *Making Sense of Global Standards*. INEF Report 58. Duisburg: Institute for Development and Peace, University of Duisburg.

Newell, Peter. 2003. Globalization and the Governance of Biotechnology. *Global Environmental Politics* 3 (2): 56–71.

Newell, Peter, and Matthew Paterson. 1998. A Climate for Business: Global Warming, the State and Capital. *Review of International Political Economy* 5 (4): 679–704.

Norberg-Hodge, Helena, Todd Merrifield, and Steven Gorelick. 2002. *Bringing the Food Economy Home: Alternatives to Global Agri-Business*. London: Zed.

Phillips, Peter, and Robert Wolfe, eds. 2001. *Governing Food: Science, Safety and Trade*. Montreal: McGill-Queen's University Press.

Ponte, Stefano. 2002. The "Latte Revolution"? Regulation, Markets and Consumption in the Global Coffee Chain. *World Development* 30 (7): 1099–1122.

Rosset, Peter. 2006. *Food Is Different: Why the WTO Should Get Out of Agriculture*. London: Zed.

Rowlands, Ian. 2000. Beauty and the Beast? BP's and Exxon's Positions on Global Climate Change. *Environment and Planning C: Government and Policy* 18 (3): 339–354.

Scharpf, Fritz. 1998. Demokratie in der transnationalen Politik. In Ulrich Beck, ed., *Politik der Globalisierung*. Frankfurt a. M.: Suhrkamp.

Sell, Susan. 1999. Multinational Corporations as Agents of Change: The Globalization of Intellectual Property Rights. In Claire Cutler, Virginia Haufler, and Tony Porter, eds., 1999. *Private Authority and International Affairs*, 169–197. Albany: SUNY Press.

Shiva, Vandana. 2000. *Stolen Harvest: Hijacking the Global Food Supply*. London: South End Press.

Sklair, Leslie. 2002. The Transnational Capitalist Class and Global Politics: Deconstructing the Corporate-State Connection. *International Political Science Review* 23 (2): 159–174.

Suppan, Steve. 2006. Codex Standards and Consumer Rights. *Consumer Policy Review* 16 (1): 5–13.

Tallontire, Anne. 2007. CSR and Regulation: Towards a Framework for Understanding Private Standards Initiatives in the Agri-food Chain. *Third World Quarterly* 28 (4): 775–791.

Weis, Tony. 2007. *The Global Food Economy: The Battle for the Future of Farming*. London: Zed.

World Bank. 2007. *World Development Report 2008: Agriculture for Development*. Washington, DC: World Bank.

Zerbe, Noah. 2004. Feeding the Famine? American Food Aid and the GMO Debate in Southern Africa. *Food Policy* 29 (6): 593–608.

I

Corporate Power in International Retail and Trade Governance

2

Retail Power, Private Standards, and Sustainability in the Global Food System

Doris Fuchs, Agni Kalfagianni, and Maarten Arentsen

Our objective in this chapter is to delineate the increasing political power of retail corporations in the food sector and to investigate its implications for environmental and social sustainability as well as food safety. In that way, we aim to highlight one of the most important current developments in global food governance and to provide a basis for the development of adequate political strategies for remedy and mitigation of potential negative effects for society. We pursue our objectives by employing a critical approach to global governance in the food system, highlighting in particular the structural and discursive facets of increasing retail power (Fuchs 2005, 2007).[1]

The last decades have witnessed a trend toward increasing capital concentration in the retail food sector. This growth in market power of a few dominant retail chains is associated with a similar increase in their political power. The latter can be noticed in the growth of the structural and discursive facets of the political power of retail corporations in particular. Market concentration, in combination with an expansion in the development of private governance institutions, specifically private food standards, reflect the growth in the structural power of retailers. At the same time, the framing of policy issues and actors in the food chain, especially retail corporations' own identity via public relations campaigns and related media activities, signals the increasing exercise of discursive power by retail corporations.

These developments, in particular the nature and consequences of private governance developed by retailers and their organizations, have crucial implications for sustainability. While promising improved food safety, private standards have ambivalent effects on the environmental and social sustainability of the global food system. Specifically, we argue that some improvements in environmental as well as nutritional and safety characteristics of food products may be expected from privately set food standards. However, these improvements will likely not go as far as retailers suggest and as consumers expect them to go. After all, conflicting interests and information asymmetries between retailers and consumers tend to weaken the stringency and effectiveness of such standards.

More importantly, however, significant detrimental effects may arise from increasing retail impact on the social sustainability of the global food system. These effects result from the marginalization of farmers from the developing countries in global markets governed by private food standards. These farmers often have to comply with private standards requiring the establishment of sophisticated and expensive systems of implementation and control (see Scott, Vandergeest, and Young, chapter 3, this volume). To stay competitive, moreover, they have to supply large volumes of food (and/or feed) per client and transaction. As a result, smallholder farmers with small economies of scale, poor access to the market, and limited investments in inputs or infrastructure often are squeezed out (Brown and Sander 2007). Similar observations exist for small, local retailers who are often squeezed out of the markets in developing and developed countries as well (Dries, Reardon, and Swinnen 2004).

What are the implications of these developments for policy and politics? First, we highlight the need for stringent political regulation of retail power in the food sector, as a means to reestablish some balance in power between business and civil society interests in global food governance. Second, we argue for a public regulatory framework for private standards, ensuring transparency, participation, and accountability in the development of standards in general, as well as the inclusion of social criteria in these standards in particular.

Based on the theoretical framework outlined in the introductory chapter, consisting of an analytic governance approach in combination with a critical power-theoretic perspective, we will now analyze the increasing power of retail corporations in global food governance and its implications for sustainability. The next section points out indicators of the increasing economic and political power of retailers. Then we delineate the implications for sustainability. The chapter's final section summarizes our findings and considers their implications for research and policy.

Expanding Retail Power

We can currently recognize an ongoing change in global food governance, at the core of which is the increasing power of retail corporations (see also Smythe, chapter 4, as well as Scott, Vandergeest, and Young, chapter 3, this volume). Aside from the favorable position of the food retailers resulting from their proximity to the customer, the convergence of three trends has led to this development. The first of these trends is the development of an oligopoly (Burch and Lawrence 2005; Konefal, Mascarenhas, and Hatanaka 2005). For a long time, there was less corporate concentration in the retail food sector than in other parts of the supply chain.[2] That has changed rapidly in the last ten to fifteen years. At the moment, we can recognize ten large internationally operating supermarket chains whose market share has constantly increased in the last two decades (see below). As we show in the following paragraphs, retailers are able to exercise significant structural and market power as a result.

Turning to the second trend, a new form of retail company has developed, characterized by the control of the product chain from farm to shelf. The underlying complex logistical task is made possible, among other things, by new technologies of supply-chain management, with which shipments are tracked by GPS and deliveries are handled in short time windows defined by the minute (Burch and Lawrence 2005).

Finally, the third trend is the development of competition that is not only based on price but also on quality (Konefal, Mascarenhas, and Hatanaka 2005). Food scandals and an increased health awareness combined with shrinking time budgets of consumers in the North have led to the emergence of new markets. The sector has enjoyed sharp increases in revenues in the luxury, organic, and health food segments. Although these markets are still referred to as "niche markets," they are the markets where most of the money will be earned in the near future. At the same time, however, discount retailers are also enjoying a boom in developed markets on the basis of cheap bulk foods. Experts expect this to be true soon for the developing markets of Asia and the Pacific, too, because several US supermarket chains are pursuing aggressive expansion in this region (Datamonitor 2006). Thus, the large retail chains have responded in a double fashion to the changing competitive landscape, with some opting for provision of expensive, high-quality foods and some others for cheaper, low-quality bulk foods.

What can we say about the power of these retail corporations in the food governance regime? Let us start with their economic power. The global retail food industry has been growing continually over the past five years and is expected to grow further at rates of from 3 to 4 percent (see figure 2.1). Europe is the largest market in retail food, generating US$1,102.3 billion in 2005, equivalent to 37.6 percent of the global industry's value. However, the Asia-Pacific market is only slightly smaller, generating US$1,002.5 billion or 34.2 percent of the global retail food industry, and growing at a faster rate than the European market, predominately due to the rapid expansion of retail food in China and India.

Within Europe and the United States, retail concentration is particularly staggering and has increased notably over the past decade.[3] Retailers purchase through just 110 buying desks,[4] which act as intermediaries between 3.2 million farmers and the consumer (MacMillan 2005). In the United States, the five largest supermarket chains have more than doubled their market share between 1992 and 2000 (Konefal, Mascarenhas, and Hatanaka 2005). Concentration, however, is also high in developing countries. Reardon, Timmer, and Berdegué (2004) report that in Latin America the top five chains per country control 65 percent of the supermarket sector. Most of the acquisitions of smaller chains and independents are foreign and taking place via foreign direct investment (FDI),

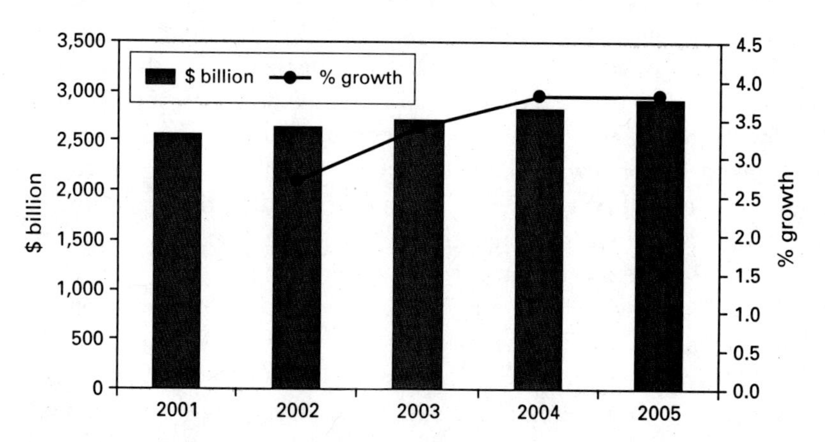

Figure 2.1
Global food retail industry value: US$ billion, 2001–2005. *Source:* Own representation on the basis of data provided by Datamonitor 2006. Global Food Retail. Industry Profile. Reference code 0199-2058. www.datamonitor.com

while some are made by larger domestic chains (Reardon, Timmer, and Berdegué 2004).[5]

In sum, a global trend toward retail concentration is evident. As figure 2.2 shows, the leading revenue source for the global retail food industry is the supermarket segment, which generated total revenues of $1,179.9 billion in 2005, equivalent to 40.3 percent of the overall market value. In comparison, the specialized-foods segment was worth $392.1 billion, which represented 13.4 percent of the market value share. For the future, economists predict the existence of six large supermarket chains that will dominate global markets and whose representatives will buy on site and distribute the products to their stores around the world through global networks (MacMillan 2005).

Currently, most of the leading global retail food chains are of European or US origin (see figure 2.3). The US-based Wal-Mart is by far the largest retailer. It recorded total revenues of $312.4 billion during the fiscal year ending in January 2006, an increase of 9.5 percent over fiscal year 2005 (Datamonitor 2006). Carrefour, the biggest European retailer and second in global rank, recorded revenues of $99.1 billion during the fiscal year ending in December 2005, an increase of 12 percent over 2004 (Datamonitor 2006, 13).

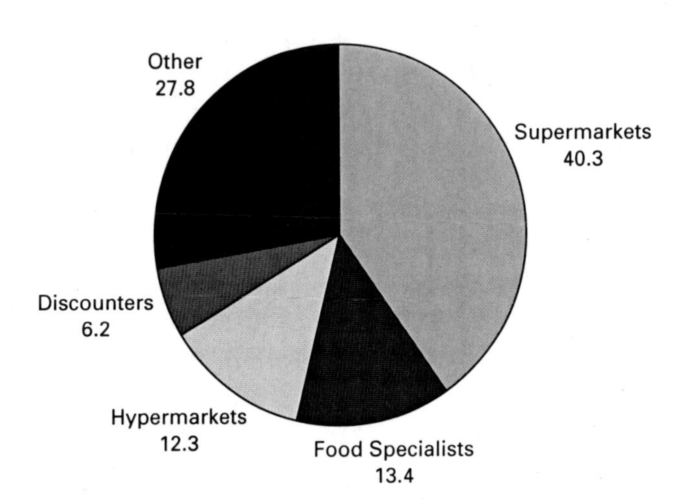

Figure 2.2
Global food retail industry segmentation: percent share, by value, 2005.[13] *Source:* Own representation on the basis of data provided by Datamonitor 2006. Global Food Retail. Industry Profile. Reference code 0199-2058. www.datamonitor.com

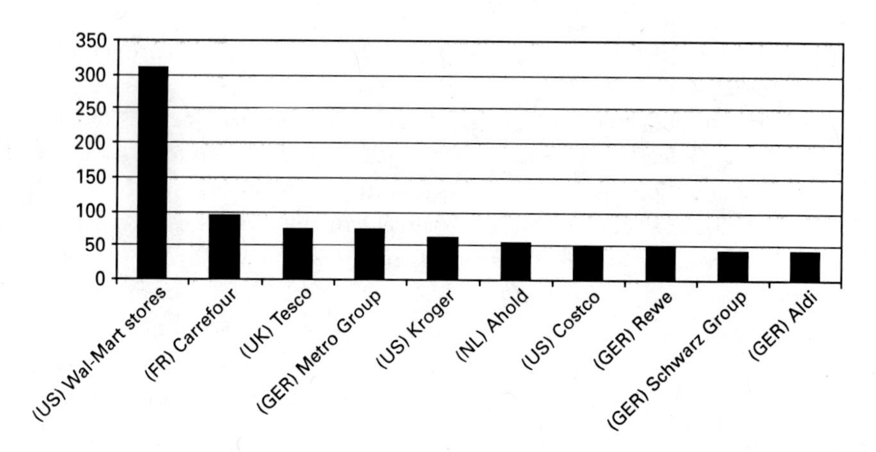

Figure 2.3
Top ten retailers 2006 annual sales in US $ billions. *Source:* Own representation on the basis of data provided by the Web site supermarket news (http://supermarketnews.com)

The market power reflected in these numbers and trends, first and foremost, implies the ability of these retail chains to put pressure on suppliers in terms of prices (oligopsony power). Importantly, the ability to exercise this pressure today reaches all the way to farmers in industrialized as well as developing countries (see also Scott, Vandergeest, and Young, chapter 3, this volume). In Norway, for instance, where four grocery chains (Norgesgruppen, Ica Norge, Coop Norge, and Reitan Narvesen) control almost 100 percent of the groceries sold, studies have documented a dramatic drop in farmers' selling power (OECD 2005). Moreover, the large grocery chains prefer selling and distributing the premium national and international brands, making it extremely difficult for small or local farmers with lesser-known brands to get their products into the grocery chains. In other words, the position in the product chain allows retail corporations to exercise significant structural power in world and domestic markets.

Importantly, however, the structural power of retail corporations today reaches beyond the question of market power. This structural power is also reflected in the development of rule-setting power—that is, with respect to private standards such as those for food safety (Fuchs 2005, 2007). Standards are defined as rules of measurement established by regulation or authority (Jones and Hill 1994), and a distinction can be made between product and process standards.[6] Most retail standards aim to ensure quality and safety at all levels of the food chain, but some also extend their requirements to environmental and social responsibility. Examples of environmental requirements include pollution prevention, use of energy, water, and other natural resources, recycling and reuse of material, and emissions. Examples of social activities can include corporate social responsibility policies and labor standards, as well as a focus on health aspects of food products.

After a number of food crises in the food chain in recent years, such as salmonella, dioxins, and BSE (bovine spongiform encephalopathy), the emphasis on safety has increased. As a result HACCP (Hazard Analysis of Critical Control Points) systems for safety management, traceability, and food safety schemes have been introduced. While HACCP is an instrument for preventing safety risks from emerging in the food chain, traceability is an instrument for rapidly identifying the source of such risks if they do appear. As such, it aims to improve food safety, reduce the risk of liability claims, and improve recall efficiency (reduce costs and enhance control of livestock epidemics) and consumer trust in the safety of food (Meuwissen et al. 2003).

Some supermarket organizations have generated their own quality assurance and safety schemes, including unexpected inspections at farms, gardens, and plants (e.g., Albert Heijn in the Netherlands, Tesco and Sainsbury in the United Kingdom, or French Carrefour with its Seal of Meat Quality and Safety). Most commonly, however, retailers develop standards collectively in order to strengthen their structural power and induce supplier participation. When all retailers endorse the same standards, they limit suppliers' market choices; in other words, the latter simply have to accept them (Busch 2000). Examples of such standards are the British Retail Consortium (BRC), the Global Food Safety Initiative (GFSI), GlobalGAP, the International Food Standard (IFS), FOODTRACE (a traceability scheme), and the Ethical Trading Initiative (ETI). Compliance with these standards by the participating agricultural and food companies is certified through independent auditors.[7]

Let us present these standards in more detail. In 1998 the British Retail Consortium developed the BRC Food Technical Standard to evaluate manufacturers' or retailers' own brand products. It includes comprehensive norms with regard to food safety and quality schemes, product and process management, and personal hygiene of personnel (in total more than 250 requirements). The majority of UK and Scandinavian retailers will only consider business with suppliers who have gained certification to the BRC Global Standard (www.brc.org.uk). The BRC introduced the Packaging Standard in 2002, followed by the Consumer Products Standard in August 2003. Each of these standards is regularly reviewed and is revised and updated at least every three years.

The Global Food Safety Initiative, initiated in 2000 by a group of international retailers, aims to ensure consumer protection, strengthen consumer confidence, set requirements for food safety, and improve efficiency costs throughout the food chain. It has fifty-two members representing 65 percent of worldwide food retail revenue. Apart from retailers, global manufacturers such as Unilever also participate.

The Global Partnership for Good Agricultural Practice (GlobalGAP) (known as EurepGAP until 2007) was developed in 1997 by a group of retailers belonging to the Euro-retailers Produce Working Group (Eurep). Initially, it covered only fruits and vegetables, but it has been expanded to cover meat products and fish from aquaculture as well. Certification is contingent on completion and verification of a checklist that consists of 254 questions, 41 of which are considered "major musts" and 122 of which are considered "minor musts." Another 91 questions are "shoulds," which are recommended but not required practices. Major musts concern traceability and food safety, while minor musts or shoulds include some environmental and animal welfare information.

The International Food Standard is a standard developed for retailers and wholesalers to ensure the safety of own-brand products. It was initiated in 2002 by German food retailers from the HDE (Hauptverband des Deutschen Einzelhandels). In 2003, French food retailers (and wholesalers) from the FCD (Fédération des Entreprises du Commerce et de la Distribution) joined the IFS Working Group and have since contributed to the development of subsequent versions of IFS (version 4).

The European retail association Eurocommerce has initiated FOODTRACE, seeking to promote a European concerted action to develop a traceability framework for the whole food chain. The initiative aspires to create a practical framework that can be used by all actors in the chain, including at the international level, to ensure traceability throughout all stages of the chain. The proposed identification scheme is supposed to be technology independent but technologically supported, in order to enable its use in developing countries.

The Ethical Trading Initiative was formed in 1998. Its aim is to develop a baseline code of conduct covering employment conditions among companies, unions, and nongovernmental organizations (NGOs), and to examine how systems of monitoring and verification can be established. As a UK initiative, its ultimate goal is to ensure that the working conditions of workers producing for the UK market at least meet international labor standards. Scholars note that the ETI should be distinguished from fair trade or alternative trade in that it does not cover small producers alone and does not carry a specific seal of approval, although companies can advertise it if they want to. Rather, it is based on written guidelines on the basis of which companies deal with their workforce in the host country (Smith and Barrientos 2005).

What is the significance of such private food standards? To some extent, they simply seem to be a response to public requests and political pressures for more transparency in the food chain (Kalfagianni 2006). Retail standards also appear to be a response to public concerns about the quality of food and food products, and they provide a signal of quality attributes to the consumer.

It is interesting to note, however, that an international governmental organization with a mandate to develop food standards ensuring transparency and quality has existed for a long time: the Codex Alimentarius Commission, founded in 1962 by the Food and Agriculture Organization (FAO) of the United Nations and the World Health Organization (see also Smythe, chapter 4, this volume). The Codex Alimentarius Commission has developed a multitude of food safety standards since its creation (Sklair 2002). In this respect, the retail companies' development of their own standards is a curious strategy. That strategy can be explained, however, if we look at participation and transparency in the development of these new private standards.

Let us take the example of the GlobalGAP standard. Initially a retailer-dominated standard, it now allows participation by retailers and suppliers alike. Yet participants from the industrialized countries, European countries in particular, dominate. In consequence, there is a clear asymmetry between North and South and between representatives of business and of civil society interests.[8] Similar observations can be made on the other retail schemes. Both the Global Food Safety Initiative and the International Food Standard limit access to retailers, while manufacturers have an advisory role. FOODTRACE, too, is a retail-led scheme for traceability in European food chains. Finally, the Ethical Trading Initiative is in principle open to participation by all stakeholders, including suppliers and civil society representatives. Resource constraints, however, especially from developing countries, provide a fundamental obstacle to participation and North-South problems remain (Schaller 2007). Regarding transparency, information about the standards is provided on the web and certain documents are available to all. However, most of the documents related to the development and monitoring of standards are only available to members. In addition, information is only provided to the general public after decisions have been made. Finally, access to information from developing countries is also problematic, due to technological constraints in most cases.

If we ignore questions of participation and transparency for the moment, we may appreciate the development of private standards, of course. For consumers in the North, these standards promise protection for food quality and safety, and in this context, the present chapter's critical reflections on these standards (see below) may stand in stark contrast to much of what is written on private food safety standards and sustainable consumption. We have to ask, however, whether they do provide sufficient protection or rather primarily aim to preempt more stringent public regulation. Moreover, these private governance institutions have fundamental implications for social sustainability and at least in this context do not represent an unequivocally positive development. We will discuss these implications shortly.

First, however, let us say a bit more on the discursive power of retail food corporations. Food retailers exercise discursive power, for instance, by representing themselves as guardians of consumer interests, which they do with respect to both prices and quality. In that context, Dixon (2007) observes that retailers are replacing traditional authorities (government, professional bodies, the church) and promote their own authority based on charisma and claims to expertise, often gained through third party association. In their representation as health authorities, for instance, supermarkets benefit from their alliance with dieticians. Furthermore, they emphasize their efficiency as market actors in production and distribution, but, more importantly, also in the design and monitoring of standards (Konefal, Mascarenhas, and Hatanaka 2005). The core argument concerning standards is that public actors act too slowly and do not have the necessary expertise to set the most efficient standards. This is a well-known argument made in many other business sectors today.

The exercise of discursive power takes place via advertising and public relations (PR) campaigns. Private standards form an important part of such PR strategies as well. While the standard guidelines are not communicated directly to consumers and are not included in product labeling, important communication takes place via retail brands, in an effort to create a loyal and stable customer base (Burch and Lawrence 2005; Codron, Siriex, and Reardon 2006). In this context, the existence of private standards incorporated in retail brands frames and signals quality assurance, as well as aiming to establish and maintain the legitimacy of the retailers as major actors in food governance. Moreover, retailers increasingly market themselves as promoters of human health via the sale of special foods. Campina, for example, highlights its development of a protein fraction that has the ability to control a specific substance in the human body (ACE), which helps to keep blood pressure at a "healthy" level (www.campina.com). The actual nutritional benefit of functional foods remains a controversial topic among scientists and regulators, of course (Lawrence and Rayner 1998; Roberfroid 2000, Katan and de Roos 2003).

The extent to which the selling of the "quality claim" is successful can be seen in the increasing market share of retail brands. Studies report that in the United States, one of every five products bought in supermarkets is of a retail store brand and further growth is expected in the future (Nemeth-Ek 2000). A similar trend is observed in several European countries. More specifically, the United Kingdom is the leader in retail brand sales with a 44.7 percent share (by volume) of the retail market, followed by Belgium with 34.8 percent, Germany with 33.4 percent, the Netherlands with 21.1 percent, and France with 22.2 percent. Further, experts expect global retail brands to achieve dominance in the food industry in the years ahead, with the biggest retailers being at the forefront of these changes (AgraEurope 2002).[9] It is noteworthy in this context that private standards are not only a manifestation of retailers' structural and discursive power, but also a means of extending it. Finally, retail companies' media presence is fundamental to their discursive power, both as a medium for constantly communicating with consumers and for adequately presenting and framing themselves.

However, there are also challenges to retailers' claims to authority. Research reveals, for example, that consumers are not always loyal to retail brands and chains, nor to the industrial food system or to the institution of the supermarket (Dixon 2007). Put the other way around, retailers are sensitive to the demands of social activists and movements for information over credence goods[10] (Hatanaka, Bain, and Busch 2005; Santoro 2003). Given the fierce competition in the sector, any negative publicity has the potential to damage sales and thereby negatively affect their economic bottom line. These social movements are often involved in the public shaming and stigmatizing of corporations for their undesirable behavior. An example can illustrate this point. In the late 1990s a series of media exposés on British television uncovered substandard labor practices among African fresh-produce suppliers selling to high-profile UK companies (Dolan 2005). More specifically, the BBC documentary *Modern Times* revealed Tesco's alleged exploitation of Zimbabwean farmworkers on prime-time television. In parallel, NGO campaigns also highlighted the harmful practices of supermarket suppliers, including the imposition of excessive working hours, sexual harassment, and exposure to pesticides on horticulture farms in Kenya (Dolan 2005). The combination of media and campaign exposure resulted in market losses for UK retailers, especially for the market segment of Kenyan vegetables in which UK supermarkets enjoyed unrivaled power. Moreover, it prompted the quick response of the ETI, which sent delegations to Nairobi to investigate the allegations of worker exploitation and pressed for more on-site social audits (Dolan 2005, 423). Such examples demonstrate the vulnerability of retailers' legitimacy and authority as political actors in the food chain and reveal that their discursive power can be contested.

Talking about rule-setting power and discursive power in retail food automatically leads to the question of consumer power. Indeed, in the political debate, both business and government actors frequently emphasize that it is the consumer who makes the importance choices. Of course, consumers do have some power. In their discussion of battles around the definition of the quality of specific food products and systems of food production and consumption, for instance, Campbell and Le Heron (2007) show how commercial processors and retailers endorsed the values of the kosher food community and organic agriculture to make profits by serving that high-value market niche.

However, this power is not as extensive as the term *consumer sovereignty*—generally mentioned in this context—suggests. Among other things, consumer power is limited by the active manipulation of demand through business actors (see media presence) and choice editing (the selection and selective presentation of goods on offer), by information asymmetries concerning products, production processes, and other actors in the supply chain, and by the high transaction costs that "political" consumers face. After all, the exercise of market power by consumers requires not merely a single consumer's decision, but the mobilization of thousands of consumers. Consumers certainly bear responsibility for the economic, social, and ethical implications of their consumption decisions. Yet, they clearly are in a weaker position within the political framework of the food governance regime than the retail industry is (see also Kinsey 2003; Lang 2003). This is also reflected in the ability of retailers to reinterpret the nature of the consumer demand for certain products and production methods. For instance, retailers have succeeded in rebranding kosher food as reflecting less of a religious preference and more of an emphasis on quality, craft, care in production, and ethical eating (Campbell and Le Heron 2007, 137). The retail commercialization of organic products has also led to a backlash against some consumer preferences (see also Scott, Vandergeest, and Young, chapter 3, this volume).

Implications of Retail Power for Sustainability

The developments in structural and to a lesser extent in discursive retail power are manifested in the promotion of private rules and standards of production with which suppliers now have to comply. Despite the aforementioned strategic reasons for which retail companies may promote private rules, the standards themselves are frequently regarded as a positive driving force toward an improvement of certain sustainability aspects of the food system. Specifically, the provision and implementation of more stringent standards regarding food safety and quality and inclusion of environmental criteria appear to be a promising development for the future sustainability of the agricultural and food sector. In this section, we assess the impact of private standards on food quality and safety, as well as on the environment and society as a whole. We also point to a few troubling issues.

Food Quality/Food Safety

The notion of food quality currently promoted by retailers refers to attributes such as appearance, cleanliness, and taste, while food safety encompasses attributes like levels of pesticide or artificial hormone residue, microbial presence in food, and so on (Reardon and Farina 2002). Treating animals humanely and originating in certain locations can also be considered quality attributes by some consumers and buyers in the food chain (Northen 2001). Environmental sustainability and fairness can be seen as elements of quality as well. Quality, then, is not an absolute and can be defined as meeting agreed-on requirements (Holleran, Bredahl, and Zaibet 1999). At the moment, however, the requirements imposed by retail standards mostly cover elements of safety and appearance rather than environmental and social sustainability.

Scholars observe that private standards and other forms of supply-chain oversight of food safety and quality are rapidly increasing not only in developed but also in developing countries (Jaffee and Henson 2004; Reardon and Berdegué 2002). Another prominent food safety and quality standard, besides the Global Food Safety Initiative and the International Food Standard mentioned above, is the HACCP, which has become normative in the food sector. It is recommended by the Codex and required by many governments (Fulponi 2006). Moreover, some private safety and quality standards are embedded in voluntary public standards at the national and/or international levels (ISO 9000) (Henson and Reardon 2005).

With the adoption of more stringent standards—both public and private—of food production and traceability schemes, the safety of the food chain has improved. In the Netherlands, for instance, a number of safety problems have been averted (dioxin in the food supply in 2004 and dioxin in animal feed in 2005). Similar incidents have been reported in Sweden and other countries. Because the reputation of retail firms depends to a great extent on food safety, they have a strong incentive to create, maintain, and expand their practices in that area.

Quality has not necessarily improved commensurately, however. Products may look better, but they do not necessarily taste better. This is a result of the economics of the production and distribution processes taking place within the global context of transnational chains and distribution networks. Fruits and vegetables, for instance, are picked early due to the long distances they usually have to be transported, so they need to be artificially gassed to ripen. Moreover, varieties are selected that can survive the harvest and shipping process and have a long shelf life. As a result, the produce ends up "relatively tasteless, nutritionally weak" (Robison 1984, 289). Nevertheless, quality standards ensure that at least some level of quality is maintained.

The implications of the spread of private food standards are less straightforward with respect to food quality in developing countries (and for the poorer sectors of society in developed countries, in fact). Optimistic observers note that higher standards for export markets can lead to spillover effects for domestic food safety in developing countries

(Jaffee and Henson 2004; Scott, Vandergeest, and Young, chapter 3, this volume). In addition, they point out that an emerging new consumer class in developing countries is likely to demand higher food quality. Van der Grijp, Marsden, and Cavalcanti (2005) report, however, that the new retail standards are leading to an increasing gap in quality between export and domestic food products.

Environmental Standards

Environmental and social standards indicate the "goodness" of production (Konefal, Mascarenhas, and Hatanaka 2005). Here, process standards are more important than product standards. Organic production standards, for instance, include prohibitions against conventional pesticides, artificial fertilizers, irradiation and food additives, or antibiotics and growth hormones for animals. Organic farming uses the environment's own systems for controlling pests and diseases in growing crops and rearing livestock. With regard to livestock, organic farming places particular emphasis on animal welfare and the use of natural foodstuffs (European Commission Staff Working Paper 2002). Environmental and social standards play a marginal role within the mainstream retail standards, but retailers are increasingly under pressure to improve at least their environmental performance.

Most of the environmental standards required by retailers are imposed through farm-practice schemes, such as Good Agricultural Practices (e.g., GlobalGAP) (Baines 2005; OECD 2006). Some standards also cover manufacturing processes and packaging, which are part of Good Manufacturing Practices (GMP) (OECD 2006). Regarding GMP, many supermarkets have recycling programs for waste reduction in packaging. Likewise, they frequently have on-site recycling facilities where consumers can return plastic bottles, glass, cans, paper products, small electronic devices, and batteries. Moreover, many supermarkets operate under ISO 14000 standards, but analysts estimate that these standards are quite minimal and have limited objectives (Clapp 2004).

Likewise, the environmental benefits from GAP are modest at the moment. The mainstream farm standards, promoted by retailers under the British Farm Standard, for instance, have been shown to address selected elements of environmental protection with respect to crop and livestock schemes only. There are examples, however, with more ambitious goals, such as Farm Biodiversity Action Plans (e.g., operated by Sainsbury's in cooperation with premium fresh produce suppliers and livestock producers), conservation plans linked to the Farming and Wildlife Advisory Group (FWAG) (Tesco's "Nature's Choice"), and the development of additional audit requirements for the Assured Produce Scheme linked to the LEAF audit (Waitrose's LEAF Marque Brand) (Baines 2005). Tesco's Nature's Choice Scheme, for instance, is an integrated management scheme introduced in 1992, which sets a wide range of environmental standards, including the rational use of plant protection products as well as fertilizers and manures; pollution prevention; the judicious use of energy, water, and other natural resources; careful recycling and reuse of material; and wildlife and landscape conservation and enhancement. Some 12,000 suppliers have to comply with these standards (www.tescocorporate.com). The objective of Sainsbury's Farm Biodiversity Action Plan is to promote biodiversity in farming and livestock. In April 2005, some 868 farms, representing 102,895 hectares and 3,482 numbers of wild species, were covered by the initiative (www.jsainsbury.com/files/reports/cr2005/). Such projects, however, concern only a small part of production so far and thus their impact on the overall environmental characteristics of the global food system is minimal. The Farm Biodiversity Action Plans, for instance, cover only 0.5 percent of the total agricultural land in the United Kingdom.

International standards such as GlobalGAP also pay some attention to environmental and social issues. As mentioned earlier, however, compliance with such issues is in most cases voluntary or recommended ("minor musts" or "shoulds"; www.globalgap.org). Thus noncompliance does not constitute a major threat to the supplier or reduce the incentive to "misbehave." More importantly, Van der Grijp, Marsden, and Cavalcanti (2005) show that the emphasis on various sustainability issues within the initiative has gradually decreased over time. As a result, GlobalGAP has turned into a program that is primarily focused on food safety and hygiene.

Finally, scholars observe that the environmental auditing of retailers has mainly been applied to particular goods (to compare and contrast performance against environmental standards), or for particular companies (to benchmark and improve performance), rather than to the whole sector (Lang and Barling 2007). Moreover, several other critical environmental issues are not covered by the retail standards. These include environmental externalities, the environmental costs of the physical relocation of shops (e.g., at the outskirts of a city), and the implications of environmental change for human health (see Lang and Barling 2007). It should also be noted in this context that GlobalGAP membership has greatly increased since the food safety focus has been strengthened, thus suggesting that food safety is a better vehicle for generating industrywide support than environment and worker welfare, at the moment. This could change in the future, when environmental issues as determinants of food quality play a bigger role in retail marketing strategies. The question remains, however, as to whether environmental standards will be stringent enough and cover a large enough segment of production so that environmental benefits can be considered significant. An analysis of the setting in terms of issues such as visibility, collective action problems, and opportunity for contestation suggests that the effectiveness of private standards with respect to environmental issues is likely to remain limited (Fuchs 2006). At the moment, then, we can only caution against being too optimistic with regard to the benefits of retail standards.

Social Issues

The social dimension of sustainability in the context of food governance covers a wide array of very complex issues ranging from workers' rights to migration and rural livelihoods, gender issues, and food security. In this context, private governance is intended to substitute for weak states, especially in developing countries that lack the capacity (and perhaps willingness) to provide and enforce social safety nets. Social provisions, such as those affecting worker welfare but also gender nondiscrimination and rules against sexual harassment, are included in mainstream standards (e.g., Ethical Trading Initiative) and companies' codes of conduct (e.g., Chiquita Code of Conduct), but they play a secondary role compared to the current understanding of food quality with its emphasis on food safety. Even though the significance of such provisions cannot be minimized, their scope is limited because they apply only to the regular workforce. Much of the labor force in developing countries is "flexible," working only seasonally, or "informal," comprising mostly female workers (Barrientos, Dolan, and Tallontire 2001; Dolan 2005).

Likewise, scholars observe that gender issues are insufficiently covered by mainstream standards. More specifically, these standards fail to recognize the different priorities for female workers stemming from the gendered nature of women's obligations to meet domestic commitments as well as their employment-related responsibilities (Pearson 2007). To some extent, this reflects the fact that codes are designed to ensure equal treatment of men and women rather than taking into account the issues that affect women because of their reproductive and societal roles (Prieto-Carrón 2006). Such limitations create significant constraints for women because of the importance of these roles.

Most fundamentally, private standards have several broader societal consequences that affect rural livelihoods, especially in developing countries. Small farmers and enterprises are being forced out of business by the high costs of implementing the new private standards, especially documentation and certification costs. An additional challenge is that most certifiers are not indigenous and have to come from Europe or the United Kingdom, which further increases the costs of certification (Hatanaka, Bain, and Busch 2005). During the last ten to fifteen years, the closing down of farm businesses and accelerated industry concentration has been a common trend, especially in developing countries. For example, thousands of small dairy operations have gone out of business in the past five years in the extended Mercosur area,[11] because they were unable to meet new quality and safety standards for milk and milk products that required large investments in equipment and buildings and a high level of coordination and management (Reardon et al. 2001). Similar developments have affected poultry operations in Central America (Alvarado and Charmel 2000). Experts predict that hundreds of thousands of peasants in Africa will lose their living as a result of the implementation of the GlobalGAP standards (ActionAid International 2005). And in Kenya and other major horticultural exporting countries in Sub-Saharan Africa, the market share of smallholders, which used to be the backbone of production, has reportedly declined significantly (Brown and Sander 2007). Instead, a few large exporters currently dominate the market, sourcing predominately from large-scale production units. More specifically, while in 1992 nearly 75 percent of fresh fruit and vegetables grown for export in Kenya were produced by smallholders, by 1998 the four largest exporters in Kenya derived only 18 percent of their produce from smallholders. In the same year, the five largest exporters in Zimbabwe sourced less than 6 percent of produce from smallholders (Brown and Sander 2007).

In addition, research on Brazil shows that the new regulatory conditions set by retailers are recreating and reinforcing other forms of economic and social cleavage. Some workers may benefit from new management practices assigning increased responsibility to an elite group of workers and the provision of bonus and competitive incentive packages for the "best" and most "loyal" workers. But a growing share of the population faces unemployment and risks losing their farms (Van der Grijp, Marsden, and Cavalcanti 2005).

These trends are resulting in highly uneven developments in the producing countries and regions (Van der Grijp, Marsden, and Cavalcanti 2005). At the same time, capital concentration in the retail sector and the global expansion of the operations of the large retail chains are threatening the livelihoods of smaller local retailers. With the increasing spread of the large retail chains to Eastern Europe and Asia, for instance, thousands of smaller, locally owned retail stores have been forced to close. Dries, Reardon, and Swinnen (2004, 7) report that the share of the modern retail sector (supermarkets, hypermarkets, and discount stores) in Central and Eastern European countries currently represents one-half and one-third in "first-wave" and "second-wave"[12] countries respectively of total retail after only a few years of "blindingly fast transformation." In Southeast Asia and East Asia (excluding China), these numbers represent 15–20 percent and 30 percent respectively (Reardon, Timmer, and Berdegué 2005). In China, the supermarket share of national retail sales of processed food is around 20 percent, similar to the supermarket share of overall retail food sales for Brazil and Argentina in the early 1990s. The difference is, however, that the rate of growth in the number of stores is three times faster in China than it was in Brazil and Argentina in the 1990s (Reardon, Timmer, and Berdegué 2005).

India provides a contrasting case. In their review of India's retailing sector and conditions for retail expansion, Neilson and Pritchard (2007) observe that the intrusion of global retail chains into the Indian market has met with several constraints, both from the demand and the supply side. On the demand side, they note that in contrast to China and countries in transition, where decades of state-managed urban retail had already altered grocery habits, the cultural idiosyncrasies of the new Indian middle-class consumer have not been affected yet. Instead, Indian consumers appear to be particularly attached to traditional wet markets and mobile grocers, whose attractiveness in a traditional family setting continues to be difficult to match. Indeed, the persistence of traditional family institutions in India characterized by limited freedom of movement for women outside the home slows down the transition to urban culture and its emphasis on food convenience provided by supermarkets. On the supply side, the Indian government has placed significant constraints on FDI in the retail sector due to social and political sensitivities attached to domestic Indian retailing. More specifically, India has a significant number of small stores (at least 15 million), accounting for 7 percent of employment and 10 percent of GDP (Neilson and Pritchard 2007, 228). Such stores often provide credit to the poor, thus also performing crucial informal financial services. The situation started to change in 2006, when partial liberalization took place, allowing 51 percent FDI in single-brand retailing. In response, local conglomerates, (RIL Reliance Industries Ltd) fearing foreign competition, however, took first steps to integrate Indian agriculture into retail supplier regimes leading commentators to speak of an Indian variant on the global supermarket system. Nevertheless, the impact of retail change on rural areas remains a critical question for India, too, and according to Scott, Vandergeest, and Young (chapter 3, this volume), for other parts of Asia as well. Currently, the well-being of many rural communities in India continues to hinge on the production of food for local self-sufficiency, managed within a complex regime of agricultural regulation orchestrated through national and state bureaucracies. Pro-market reforms promising to improve the economic welfare of the bulk of India's rural poor are viewed with skepticism. Likewise, fears are voiced that such reforms will increase rural landlessness and accelerate urban-rural migration, with consequent population pressures on India's already overburdened metropolitan infrastructure (Neilson and Pritchard 2007, 237).

Why should we care about small farmers and retail stores? After all, one could argue that they simply need to be of a sufficient size to be able to compete in global markets, and that that size today is rather large. However, the owners of small farms frequently do not have alternatives, because social safety nets do not exist in many developing countries and unemployment is high, not just in the countryside but also in urban areas. Accordingly, rural populations forced to abandon agriculture and move to cities do not find alternative sources of income as industrial workers, as had been the case in Europe when it industrialized. On the contrary, the social consequences of these standards carry the risk of turning these farmers into subsistence farmers, who already form the bulk of the global population suffering from hunger today.

A critical question in that respect is whether the elimination of small business operations is desirable from a food safety and quality perspective. After all, consumer anxiety over safety and quality has heightened as a result of a number of food crises (Goodman and DuPuis 2002).

Although inadequate statistics make it difficult to estimate the global incidence of foodborne diseases, and there is substantial underreporting, data from industrialized countries indicate that up to 10 percent of populations are affected annually by foodborne diseases (Saker et al. 2004). Are large suppliers better able to address the food safety and quality concerns that have become all the more easily transferable with the globalization of food trade? Initial evidence suggests that opportunities for the technological and organizational innovation necessary for food safety control are limited to large suppliers only, due to the financial constraints mentioned earlier. However, these suppliers are also more susceptible to harboring and spreading diseases because of the concentration of production. Centralized processing and mass distribution may lead to widespread dissemination of contaminated foods (Saker et al. 2004). In contrast, a supply base of numerous small-scale producers that are geographically dispersed often acts as an effective mechanism for reducing the risk of widespread crop failures due to disease—and presumably also for reducing the risk of foodborne illness that can affect consumers. Small-scale producers, moreover, have a comparative advantage in supplying high-value agricultural products—for example, when the production practices required to meet quality standards are labor intensive (e.g., fine beans and baby corn) (Henson, Masakure, and Boselie 2005). Thus, arguments linking concentration trends to improved safety are unconvincing.

Optimistic views on the implications of private standards for social sustainability exist as well. However, they have to be regarded with caution. Some observers argue, for instance, that the formal certification process benefits (some) farmers, because it opens up new opportunities in global export markets. Firms that implement standards are said to be able to increase efficiency and therefore profit rates through better intrafirm and interfirm coordination and management (Mazzocco 1996). Moreover, standards make it possible to reach more consumers by communicating and reassuring them regarding safety and quality, especially in the absence of public standards (Reardon and Farina 2002). Thus, GlobalGAP participants in Ghana report that the new standards increased their credibility in the global market and helped them reduce pesticide use (Hatanaka, Bain, and Busch 2005). However, such opportunities tend to exist only for a small subset of suppliers—those receiving capacity-building support from NGOs, development agencies, or multilateral agencies (Reardon and Farina 2002).

Likewise, more optimistic evaluations point out that new stringent food standards could provide an incentive to modernize production. They report that given the capacity to modernize, food standards could be a basis for competitive repositioning and enhanced export performance of developing countries (Jaffee and Henson 2004). At the moment, however, only the large retail chains and agribusinesses have the necessary financial capacity for such measures, and it is questionable whether this will change in the near or midterm future.

To summarize, retail standards have a positive effect on food safety and to some extent on quality. With regard to developing countries, however, these benefits tend to be limited to food products for export markets and potentially for the financially better off, emerging urban consumer class. Some environmental benefits can be expected from the introduction of environmental standards. But at the moment, environmental concerns play a marginal role. The area of greatest concern is the impact of retail power on the social aspects of sustainability, particularly in developing countries. The implementation of stringent retail standards has driven many farmers and local retailers out of business and led to increasing social inequality.

Conclusion

In this chapter, we have delineated facets of the increasing political and economic power of retail corporations and investigated their implications for sustainability. We have shown that retailers have acquired rule-setting power as a result of their material position within the global economy. Due to their control of networks and resources, retailers have gained the capacity to adopt, implement, and enforce private rules (standards), which then take on an obligatory quality. In other words, retailers have been able to punish and reward (countries and) suppliers for their choice of rules and standards, thereby limiting their choices. In this context, retailers have benefited from their acquisition of political legitimacy and therefore discursive power, which results from their alleged expertise and efficiency in relation to public actors; they have then used their rule-setting activities to enhance and maintain this power. Finally, retail companies' media presence has been fundamental to the expansion of both their structural and discursive power.

The impact on sustainability of growing retail power has been both positive and negative so far. On the one hand, increasing retail power can be linked to some improvements in safety and quality. Some environmental benefits have been gained from the introduction of private standards as well, but at the moment, important issues are excluded and implementation is weak, and so the impact on sustainability is mixed. But the situation could change in the future as environmental improvement increasingly comes to be viewed as relevant to food quality.

When we consider the impact of retailers' power on the social dimensions of sustainability and particularly on farmers in developing countries, however, we can clearly see the negative consequences of private governance by retail standards. The implementation of retail standards is driving many small and medium-sized farmers out of business. Likewise, the capital concentration in the retail sector and the global expansion of the operations of the large retail chains are threatening the livelihoods of smaller local retailers in developing and developed countries. Thus, increasing retail power is also fostering increasing inequality and poverty.

What can be done to improve the current situation? First, our analysis highlights the need to prevent the emergence of monopsony and oligopsony power, because such power implies asymmetries in both economic and political power that are detrimental to sustainable development. The European Commission, in this context, has stopped mergers between companies in cases where there were justified fears of monopsony power (OECD 2006). Yet, often the problem is to find proof of unfair seller (farmer) prices, especially when farmers are located in developing countries. A partial solution could be the development of farmer cooperatives, as well as investment in collective action and networking and other forms of strategic alliances that help balance the power exercised by retail corporations (Cook and Iliopoulos 2000; Fulton 2001; Johnson and Berdegué 2004; Levins 2002). Bain and Busch (2004) refer to the successful formation of two processing cooperatives in Michigan to meet the challenge of increasingly rigid private standards. Henson, Masakure, and Boselie (2005) also present the example of Hortico Fresh Produce Ltd., one of the largest fresh-produce exporters in Zimbabwe with 4,000 small suppliers, which have successfully negotiated with big retailers regarding specific elements of standards in order to find ways to comply at minimal cost. This approach, however, needs to accommodate two concerns. First, the desirability of farmer cooperatives depends on their size. A trend toward farmer monopoly (or oligopoly) would be as undesirable as a trend toward retailer monopsony (or oligopsony). Second, collective action, through unionization for instance, has declined considerably across the world. It has apparently dropped especially sharply in Latin America, with decreased levels of unionization by almost 50 percent in Argentina, Colombia, Peru, and Venezuela since 1980 (Sabatini and Farnsworth 2006). A relevant question in this context is whether the "left turn" in Latin America will reverse this situation or whether its leadership will prove merely populist and hence, unequipped to accommodate farmers' and, in general, workers' concerns.

In parallel, standards need to start incorporating social criteria. To some extent, it would be in the interest of retailers themselves to make social criteria an integral part of their standards. A healthy environment and a healthy workforce, with essential rights and a decent quality of life, would appear to be prerequisites to retailers' long-term economic success. Cynical observers might argue, of course, that such standards would only include weak requirements and seek to legitimize existing retail practices. Yet, they would also provide workers and farmers with rights and companies with obligations. As soon as a code of conduct or a private standard with social criteria is there, it can always be challenged and improved.

If we do not trust retailers to develop standards with sufficiently stringent social criteria in the near future, we will have to rely on public actors. An obvious candidate with sufficient weight in the global food market that can influence decisions on agriculture and food issues at the international level is the EU. The EU has proven its capacity in this arena with the adoption of the General Food Law (178/2002/EC), which imposed concrete demands on traceability and food safety. Even though the final responsibility for the development of such schemes lies with private actors, they are accountable to the EU institutions. EU pressure on retailers to adopt certain standards could also lead to the endorsement of such standards in national legislation abroad.

Such a move by the EU, however, could be considered an effort to export its own norms to its trading partners, reflecting a new form of colonialism. Besides, the EU is only one of the major public actors with the ability to influence food standards at the global level. Coordinated efforts would also require consensus within the Codex, something that has proven extremely difficult so far, especially in the case of labeling genetically modified food (see Smythe, chapter 4, this volume). That is why the participation of NGOs and farmers from developing countries in the creation of social codes of conduct or social criteria in private standards is necessary. Participation by different actors could also allow the development of different instruments and regulatory arrangements better suited to addressing the complex aspects of food governance (see also Martinez et al. 2007). After all, reliance on "single-instrument" approaches is often misguided because none are flexible and resilient enough to be able to successfully address all problems in all contexts (Gunningham and Sinclair 1999). Yet, a great difficulty in that respect is that the countries in which agricultural practices require the most social provisions frequently lack the institutional environment to foster such horizontal negotiations. Accordingly, effective local participation will require special efforts and capacity-building measures by the EU, FAO, and international activist groups.

Such activities by public actors, then, could create a framework ensuring the adequate performance of private standards. Specifically, such a framework could foster the creation of food standards that simultaneously promote the different dimensions of sustainability: food safety in the narrow sense, environmental quality, and social quality. Without such activities, the growth and concentration of retail power described in this chapter and the associated expansion of the role of private food standards can be expected to have a mixed outcome for the sustainability of the global agrifood system.

Acknowledgments

We would like to thank three anonymous reviewers, and, in particular, Jennifer Clapp and the participants in the Waterloo workshop on "Corporate Power in Global Agrifood Governance" for helpful comments on this chapter. Moreover, the authors are grateful for the research assistance of Philipp Forstner and Frederike Boll.

Notes

1. We concentrate on structural and discursive power because these are the areas in which we find the most interesting developments. We do not mean to imply that instrumental power does not constitute an important part of retailers' political power.

2. Concentration of the processing and manufacturing sectors started much earlier than concentration of the retail sector. Food giants, such as Nestlé and Unilever in Europe and Kraft and General Foods in the United States, were the result of mergers and acquisitions that began already in the 1980s (Lang 2003). At that time the retail sector was still characterized by relatively smaller shops.

3. Concentration is measured as the percentage of five-firm control of the retail market in grocery and daily household goods.

4. Buying desks carry out the purchasing for retailers. Today, they usually do this for groups or consortia of retailers at the same time.

5. The same source reports that in the first eight months of 2002, for example, five global retailers (Tesco, Carrefour, Ahold, Makro, and Food Lion) spent US$120 million in Thailand and Wal-Mart spent $660 million in Mexico to build new stores (p. 9).

6. Product standards refer to various characteristics of the product itself, like quality or safety. Process standards refer to the processing of raw products into intermediate or final goods. They specify the characteristics that the processes are expected to have, either to produce products with specific attributes (organic, safe) or to create and maintain certain conditions for the environment, workers, and so on (Reardon et al. 2001).

7. The ETI as an organization, however, does not audit any companies. Instead, its members submit annual reports to the board, based on audits conducted either by third parties or in-house auditors.

8. These asymmetries in participation are all the more noteworthy, because even the Codex had been widely criticized for allowing too much corporate influence.

9. The fastest growth with 6.7 percent in volume and nearly 5 percent in value has been reported in France, and growth exceeded 4 percent in Germany and 3 percent in Belgium (Nemeth-Ek 2000).

10. *Credence goods* are those whose quality cannot be directly experienced by the consumer without additional information—for example, goods produced by environmentally friendly methods or by ethical codes of conduct.

11. Mercosur is the South American Trade Block, with Brazil, Argentina, Uruguay and Paraguay as core members, and Chile and Bolivia as associate members.

12. First-wave countries are defined as those "starting" globalization around 1996, such as the Czech Republic, Hungary, and Poland. Second-wave countries are defined as those "starting" globalization in the late 1990s, such as Croatia, Romania, and Bulgaria.

13. A *hypermarket* is a very large self-service store that sells products usually sold in department stores as well as those sold in supermarkets—for example, clothes, hardware, electrical goods, and food. *Food specialists* are stores which sell specific food products, such as organic and fair trade products, but also delicatessen or pet food, for instance.

References

ActionAid International. 2005. *Power Hungry: Six Reasons to Regulate Global Food Corporations*. London: ActionAid International. www.actionaid.org.uk/_content/documents/power_hungry.pdf.

AgraEurope. 2002. The Shape of Things to Come—Food Industry Report, Global Brands to Dominate Food Industry over Next Year. *Eurofood*, February 28.

Alvarado, Irene, and Kiupssy Charmel. 2000. Crecimiento de los canales de distribucion de productos agricolas y sus efectos en el sector rural de America Central. Paper presented at the International Workshop Concentration in the Processing and Retails Segments of the Agrifood System in Latin America: Effects on the Rural Poor. Santiago, Chile, November.

Bain, Carmen, and Lawrence Busch. 2004. *Standards and Strategies in the Michigan Blueberry Industry*. Michigan Agricultural Experiment Station Report 585. East Lansing: Michigan State University.

Baines, Richard. 2005. *Private Sector Environment Standards: Impact on Ecological Performance and International Competitiveness of UK Agriculture*. Final Report RES-224–25–0036. Newcastle: Rural Economy and Land Use Programme, School of Agriculture, Food and Rural Development, University of Newcastle. http://www.relu.ac.uk/research/projects/Baines.htm.

Barrientos, Stephanie, Catherine Dolan, and Anne Tallontire. 2001. *The Gender Dilemma in Ethical Trade*. NRI Working Paper No. 2624. Chatham: Natural Resources Institute.

Brown, Oli, and Christina Sander. 2007. *Supermarket Buying Power: Global Supply Chains and Smallholder Farmers*. Winnipeg: IISD. http://www.tradeknowledgenetwork.net/pdf/tkn_supermarket.pdf.

Burch, David, and Geoffrey Lawrence. 2005. Supermarket Own Brands, Supply Chains and the Transformation of the Agri-Food System. *International Journal of Sociology of Agriculture and Food* 13 (1): 1–18.

Busch, Lawrence. 2000. The Moral Economy of Grades and Standards. *Journal of Rural Studies* 16 (3): 273–283.

Campbell, Hugh, and Richard Le Heron. 2007. Supermarkets, Producers and Audit Technologies: The Constitutive Micropolitics of Food, Legitimacy and Governance. In David Burch and Geoffrey Lawrence, eds., *Supermarkets and Agri-food Supply Chains*, 131–153. Cheltenham: Edward Elgar.

Clapp, Jennifer. 2004. The Privatization of Global Environmental Governance: ISO 14000 and the Developing World. In David Levy and Peter Newell, eds., *The Business of Global Environmental Governance*, 223–248. Cambridge, MA: MIT Press.

Codron, Jean-Marie, Lucie Siriex, and Thomas Reardon. 2006. Social and Environmental Attributes of Food Products in an Emerging Mass Market: Challenges of Signalling and Consumer Perception with European Illustrations. *Agriculture and Human Values* 23 (3): 283–297.

Cook, Michael L., and Constantine Iliopoulos. 2000. Ill-defined Property Rights in Collective Action: The Case of US Agricultural Cooperatives. In Claude Menard, ed., *Institutions, Contracts and Organisations: Perspectives from New Institutional Economics*, 335–348. Northampton: Edward Elgar.

Datamonitor. 2006. *Global Food Retail: Industry Profile.* Reference code 0199–2058. www.datamonitor.com.

Dixon, Jane. 2007. Supermarkets as New Food Authorities. In David Burch and Geoffrey Lawrence, eds., *Supermarkets and Agri-food Supply Chains*, 29–50. Cheltenham: Edward Elgar.

Dolan, Catherine S. 2005. Benevolent Intent? The Development Encounter in Kenya's Horticulture Industry. *Journal of Asian and African Studies* 40 (6): 411–437.

Dries, Liesbeth, Thomas Reardon, and Johan F. M. Swinnen. 2004. The Rapid Rise of Supermarkets in Central and Eastern Europe: Implications for the Agrifood Sector and Rural Development. *Development Policy Review* 22 (5): 1–32.

European Commission Staff Working Paper. 2002. *Analysis of the Possibility of a European Action Plan for Organic Food and Farming.* Brussels: Commission of the European Communities. SEC(2002)1368, 12 December.

Fuchs, Doris. 2005. Commanding Heights? The Strength and Fragility of Business Power in Global Politics. *Millennium* 33 (3): 771–803.

Fuchs, Doris. 2006. Transnational Corporations and Global Governance: The Effectiveness of Private Governance. In Stefan Schirm, ed., *Globalization: State of the Art of Research and Perspectives*, 122–142. London: Routledge.

Fuchs, Doris. 2007. *Business Power in Global Governance.* Boulder: Lynne Rienner.

Fulponi, Linda. 2006. Private Voluntary Standards in the Food System: The Perspective of Major Retailers in OECD Countries. *Food Policy* 31 (1): 1–13.

Fulton, Murray. 2001. Traditional versus New Generation Cooperatives. In Christopher D. Merrett and Norman Walzer, eds., *A Cooperative Approach to Local Economic Development*, 11–24. Westport: Quorum Books.

Goodman, David, and E. Melanie DuPuis. 2002. Knowing Food and Growing Food: Beyond the Production-Consumption Debate in the Sociology of Agriculture. *Sociologia Ruralis* 42 (1): 5–22.

Gunningham, Neil, and Darren Sinclair. 1999. Regulatory Pluralism: Designing Policy Mixes for Environmental Protection. *Law and Policy* 21 (1): 49–76.

Hatanaka, Maki, Carmen Bain, and Lawrence Busch. 2005. Third-Party Certification in the Global Agrifood System. *Food Policy* 30 (3): 354–369.

Henson, Spencer, Oliver Masakure, and David Boselie. 2005. Private Food Safety and Quality Standards for Fresh Produce Exporters: The Case of Hortico Agrisystems, Zimbabwe. *Food Policy* 30 (4): 371–384.

Henson, Spencer, and Thomas Reardon. 2005. Private Agri-Food Standards: Implications for Food Policy and the Agri-Food System. *Food Policy* 30 (3): 241–253.

Holleran, Erin, Maury E. Bredahl, and Lokman Zaibet. 1999. Private Incentives for Adopting Food Safety and Quality Assurance. *Food Policy* 24 (6): 669–683.

Jaffee, Steven, and Spencer Henson. 2004. *Standards and Agro-Food Exports from Developing Countries: Rebalancing the Debate*. World Bank Policy Research Paper No. 3348. Washington, DC: World Bank.

Johnson, Nancy, and Julio A. Berdegué. 2004. Collective Action and Property Rights for Sustainable Development: 2020 Vision for Food, Agriculture and the Environment. In Ruth S. Meinzen-Dick and Monica Di Gregorio, eds., *Collective Action and Property Rights for Sustainable Development*. Washington, DC: International Food Policy Research Institute (IFPRI).

Jones, Eluned, and L. D. Hill. 1994. Re-engineering Marketing Policies in Food and Agriculture: Issues and Alternatives for Grain Grading Policies. In Daniel I. Padberg, ed., *Re-engineering Marketing Polices for Food and Agriculture*, 119–129. College Station: Texas A&M Food and Agricultural Marketing Consortium.

Kalfagianni, Agni. 2006. *Transparency in the Food Chain: Policies and Politics*. Twente: Twente University Press.

Katan, Martijn B., and Nicole M. de Roos. 2003. Public Health: Toward Evidence-Based Health Claims for Foods. *Science* 299 (5604): 206–207.

Kinsey, Jean. 2003. Emerging Trends in the New Food Economy: Consumers, Firms and Science. Paper presented at OECD Conference on Changing Dimensions of the Food Economy, The Hague, February 6–7.

Konefal, Jason, Michael Mascarenhas, and Maki Hatanaka. 2005. Governance in the Global Agro-Food System: Backlighting the Role of Transnational Supermarket Chains. *Agriculture and Human Values* 22 (3): 291–302.

Lang, Tim. 2003. Food Industrialisation and Food Power: Implications for Food Governance. *Development Policy Review* 21 (5–6): 555–568.

Lang, Tim. 2003. Battle of the Food Chain. *Guardian*. Saturday May 17, 2003.

Lang, Tim, and David Barling. 2007. The Environmental Impact of Supermarkets: Mapping the Terrain and Policy Problems in the UK. In David Burch and Geoffrey Lawrence, eds., *Supermarkets and Agri-Food Supply Chains*, 192–215. Cheltenham: Edward Elgar.

Lawrence, Mark, and Mike Rayner. 1998. Functional Foods and Health Claims: A Public Health Policy Perspective. *Public Health Nutrition* 1 (2): 75–82.

Levins, Richard. 2002. Collective Bargaining by Farmers: Fresh Look? *Choices* (winter 2001–2002): 15–18.

MacMillan, Tom. 2005. *Power in the Food System: Understanding Trends and Improving Accountability*. Food Ethics Council Background Paper. Brighton: Food Ethics Council.

Martinez, Marian G., Andrew Fearne, Julie A. Caswell, and Spencer Henson. 2007. Co-Regulation as a Possible Model for Food Safety Governance: Opportunities for Public-Private Partnerships. *Food Policy* 32 (4): 299–314.

Mazzocco, Michael A. 1996. HACCP as a Business Management Tool. *American Journal of Agricultural Economics* 78 (3): 770–774.

Meuwissen, Miranda, P. M. Alfons, G. J. Velthuis, Henk Hogeveen, and Ruud B. M. Huirne. 2003. Traceability and Certification in Meat Supply Chains. *Journal of Agribusiness* 21 (2): 167–181.

Neilson, Jeffrey, and Bill Pritchard. 2007. The Final Frontier? The Global Roll-Out of the Retail Revolution in India. In David Burch and Geoffrey Lawrence, eds., *Supermarkets and Agri-food Supply Chains*, 219–242. Cheltenham: Edward Elgar.

Nemeth-Ek, Maria. 2000. Private Label Brands Captivate Europe's Consumers. *AgExporter*, January 2000. http://www.fas.usda.gov/info/agexporter/2000/Jan/private.html.

Norten, J. R. 2001. Using Farm Assurance Schemes to Signal Food Safety to Multiple Retailers in the U.K. *International Food and Agribusiness Management Review* 4 (1): 37–50.

OECD. 2005. *Competition and Regulation in Agriculture: Monopsony Buying and Joint Selling*, May. http://www.oecd.org/dataoecd/7/56/35910977.pdf.

OECD. 2006. *Final Report on Private Standards and the Shaping of the Agro-Food System*. Paris: OECD.

Pearson, Ruth. 2007. Beyond Women Workers: Gendering CSR. *Third World Quarterly* 28 (4): 731–749.

Prieto-Carrón, Marina. 2006. Corporate Social Responsibility in Latin America: Chiquita, Women Banana Workers and Structural Inequalities. *Journal of Corporate Citizenship* 21: 1–10.

Reardon, Thomas, and Julio A. Berdegué. 2002. The Rapid Rise of Supermarkets in Latin America: Challenges and Opportunities for Development. *Development Policy Review* 20 (4): 371–388.

Reardon, Thomas, Jean-Marie Codron, Lawrence Busch, James Bingen, and Craig Harris. 2001. Global Change in Agrifood Grades and Standards: Agribusiness Strategic Responses in Developing Countries. *International Food and Agribusiness Management Review* 2 (3): 421–435.

Reardon, Thomas, and Elizabeth Farina. 2002. The Rise of Private Food Quality and Safety Standards: Illustrations from Brazil. *International Food and Agribusiness Management Review* 4 (4): 413–421.

Reardon, Thomas, C. Peter Timmer, and Julio A. Berdegué. 2004. The Rapid Rise of Supermarkets in Developing Countries: Induced Organisational, Institutional and Technological Change in Agrifood Systems. Paper presented at the Meetings of the International Society for New Institutional Economics, Tucson, Arizona, September.

Reardon, Thomas, C. Peter Timmer, and Julio A. Berdegué. 2005. Supermarket Expansion in Latin America and Asia: Implications for Food Marketing Systems. *New Directions in Global Food Markets*. Agriculture Information Bulletin No. (AIB794) 81: 47–61.

Roberfroid, Marcel B. 2000. Concepts and Strategy of Functional Food Science: The European Perspective. *American Journal of Clinical Nutrition* 71 (6): 1660–1664.

Robison, Wade L. 1984. Management and Ethical Decision Making. *Journal of Business Ethics* 3 (4): 287–291.

Sabatini, Christopher, and Eric Farnsworth. 2006. The Urgent Need for Labor Law Reform. *Journal of Democracy* 17 (4): 50–63.

Saker, Lance, Kelley Lee, Barbara Cannito, Anna Gilmore, and Diarmid Campbell-Lendrum. 2004. *Globalisation and Infectious Diseases: A Review of the Linkages*. Geneva: World Health Organisation. http://www.who.int/tdr/cd_publications/pdf/seb_topic3.pdf.

Santoro, Michael A. 2003. Beyond Codes of Conduct and Monitoring: An Organizational Integrity Approach to Global Labor Practices. *Human Rights Quarterly* 25 (2): 407–424.

Schaller, Susanne. 2007. *The Democratic Legitimacy of Private Governance: An Analysis of the Ethical Trading Initiative*. INEF Report 91/2007. Institute for Development and Peace. Duisburg: University of Duisburg-Essen. http://inef.uni-due.de/page/documents/Report91.pdf.

Sklair, Leslie. 2002. The Transnational Capital Class and Global Politics. Deconstructing the Corporate-State Connection. *International Political Science Review* 23 (2): 159–174.

Smith, Sally, and Stephanie Barrientos. 2005. Fair Trade and Ethical Trade: Are There Moves Towards Convergence? *Sustainable Development* 13 (3): 190–198.

Van der Grijp, Nicolien M., Terry Marsden, and Josefa S. B. Cavalcanti. 2005. European Retailers as Agents of Change towards Sustainability: The Case of Fruit Production in Brazil. *Environmental Sciences* 2 (4): 445–460.

3

Certification Standards and the Governance of Green Foods in Southeast Asia

Steffanie Scott, Peter Vandergeest, and Mary Young

With the advent of widespread ecological and socioeconomic concerns over the model of "green revolution" agriculture in the global South, social movements emerged in the 1970s and 1980s to work with farmers to develop "alternative" agricultural practices. In the case of Southeast Asia, these were the seeds of current-day organic production, which has become increasingly subject to certification in order to market products in corporate-controlled supply chains. These shifts in Southeast Asia's agrifood system have paralleled growing concerns by consumers—subsequently played up in retailer advertising—in the global North and South about the quality and safety of the food system. Discriminating consumers have increasingly demanded environmentally friendly production that is respectful of animal welfare, labor, and social standards. This in turn has led to the introduction of a plethora of ecolabels and standards, set by public- as well as private-sector agencies, including standards for organic production. Other transformations in the agrifood system in recent years include shifts in the market power of agrifood corporations from manufacturing to retailing, a stricter regulatory environment, a stronger voice of consumers and civil society, and the globalization of supply and distribution systems (Fulponi 2007).

These trends have understandably led to concerns over the corporate role in the food system, including retail power (Lang and Heasman 2004; Fuchs, Kalfagianni, and Arentsen, chapter 2, this volume), and to growing demands for transparency in food system governance (see Smythe, chapter 4, this volume). Many consumers, food activists, and academics have associated organic or similarly labeled foods with healthiness, environmental sustainability, and more broadly with a critique of conventional or industrial agriculture, as summarized by Julie Guthman (2004, 3–9; Pugliese 2001). More recently, however, agrifood scholars and activists have elaborated a scathing critique of the industrialization of organic farming (Buck, Getz, and Guthman 1997; Guthman 2004; Pollan 2006). Observers accuse the processing and transportation in the organic sector of "floating in a sinking sea of petroleum" (Pollan 2006, 184) as organic greens produced on corporate Californian farms are trucked to supermarkets in Toronto, and organic apples are shipped from New Zealand.

At the farm level, meanwhile, local ecological knowledge has been supplanted by international standards that remove organic farming from social and ecological contexts and reduce it instead to a set of techniques and documentation requirements that can be applied around the world for marketing purposes (Campbell and Stuart 2005; Vandergeest 2007; Mutersbaugh 2005a). According to Campbell and Stuart (2005, 95), "By shifting the disciplining of organic from the mutable immobile realm of co-production between the organic social movement, growers, and biophysical nature at the local level, to the wider audit disciplines of immutable mobile global standards, the chances for long term sustainable outcomes for organic production are clearly diminished." One response among those who reject these trends has been to put more emphasis on buying from local farmers, now codified in popular terms like "food miles" or the hundred-mile diet (Smith and Mackinnon 2007).

In this chapter we outline how these debates have emerged in the global South, drawing on our research in three Southeast Asian countries: Thailand, Indonesia, and Vietnam. We track the various controversies over the growth in influence of local and transnational agrifood corporations into organic agriculture. First we review the debates over the conventionalization or mainstreaming of organic production, and the centrality of certification and corporate interests in this process. Here we show that corporate demand for and involvement in the certification process for green and organic foods represents a form of structural power, which in turn shapes the certification rules and processes. We then show that corporations have taken advantage of the past efforts of social movements and nongovernmental organizations (NGOs) working with farmers to develop alternative agricultural practices. Here corporate players have used discursive strategies to capture the discourse around "safe foods," which in turn strengthens their structural power in the marketplace. In comparing the way these processes unfolded in Thailand, Indonesia, and Vietnam, we argue that emerging debates over the corporate role in organic agriculture and food marketing take different forms in the three sites depending on a number of factors. These include, among other things, the relative influence of vigorous social movements promoting alternative agriculture, the degree to which agriculture is export oriented, the relative importance of supermarkets in the domestic food market, and the support of government and donor agencies for organic production.

Corporate Power, Certification, and the Mainstreaming of Organic Agriculture

By 2009, the global market for organic foods is expected to exceed US$86 billion, up from US$33 billion in 2005 and US$23 billion in 2002 (Global Industry Analysts 2007; Sahota 2007). The Asia-Pacific region is projected to be the fastest growing market for organic foods (Global Industry Analysts 2007). The world of organic certification is increasingly complex and contested, encompassing to various degrees private, state, and local processes. The shift toward a certification regime in organic agriculture has enabled agrifood retailers to position themselves as the main providers of "sustainable" and "safe" agricultural products. Because of their centrality in the marketing of such foods, corporate players have had a degree of structural power that has enabled them to influence the direction of the certification process. This power derives not only from their large and growing share of the food retailing industry and the growth in sales of organic foods but also from their role in the establishment of private and quasi-public, quasi-private certification standards for organic foods. Their position has also enabled them to "capture" the discourse around organic agriculture to some extent, which only reinforces their structural power by enabling them to increase their market share.

Organic and related farming systems (e.g., biodynamic) emerged during early twentieth century in the global North, starting in Europe and spreading to North America, Australia, and New Zealand. Organic farming was based on a critique of environmental and human health effects of petrochemical-based agri-inputs, and instead emphasized an integrated approach to cultivating the vitality of soils, plants, animals, and human health.

Certification of organic production was a response to growing consumer interest in organically produced food during the 1960s and 1970s;

because this was an "extrinsic" quality, not visibly detectable in the product itself, consumers and retailers needed to find a way of clearly identifying foods produced by farms that met clear standards. Initially, these guarantees were often provided by social movement organizations through personalistic and local associations—producer associations, food cooperatives, cafés, and so on (e.g., see Campbell and Stuart 2005; Mutersbaugh 2005b, 397; Raynolds and Wilkinson 2007). In other words, certification is generally unnecessary in cases of direct marketing in which consumers know and trust producers. Certification tends to be introduced by agribusiness to create consumer confidence in instances where the consumer is not in direct personal contact with the producer (Johannsen, Wilhelm, and Schone 2005). In this way, certification can be understood as an artifact of an increasingly commodified agrieconomy that distances consumers from food production.

As the products of organic agriculture moved into more anonymous markets, with longer supply chains, larger organizations like the UK's Soil Association formulated general standards and inspection processes. The Soil Association began inspections and certification in 1973 (Smith 2006, 448). The 1970s also marked the formation of the International Federation of Organic Agriculture Movements (IFOAM), an umbrella organization of the major organic farming organizations of the global North; it facilitated the growth of organic certification through the formulation of its "basic standards," first published in 1980. These standards, which are continually under revision, have formed the basis for private, state, and United Nations (Codex Alimentarius) organic standards (Organic Europe, n.d.).

The certification of organic food has been given an additional push as food safety and traceability have become cornerstones of new "value-added" qualities through which agrifood corporations have sought to expand what could otherwise be a limited market growth potential in the global North, given slow-growing populations and "natural" limits on how much food any one individual can consume. Retailers are taking advantage of consumer concerns about food safety and environmental issues through niche marketing of "green" foods, the labels of which can translate into significant price premiums paid by consumers. In this way firms have exercised discursive power by shaping the discourse around organic foods and using this as a means by which to increase their share of the marketing of such products. Linking up with structural power of firms, private and commercial organic certification organizations emerged in part from the demand by supermarket chains to facilitate arrangements with producers, and from a desire to sell qualities like food safety, animal welfare, environmental protection, and worker welfare to consumers.

As organic commodities have continued to grow in economic significance, additional layers of regulation and accreditation were formed, through intersecting networks of government, intergovernmental, NGO, and private bodies. Beginning with the EU's adoption of organic agriculture standards in 1992, governments have created standards that increasingly restrict the use of the "organic" label, and have instituted accreditation of certifying bodies. The EU standards and process have been the model for many other governments. Also of growing importance are the standards adopted by the UN's Codex Alimentarius, which are intended to be used for resolving trade disputes where exporting countries claim that importing government restrictions on the use of the organic label constitute a trade barrier. Meanwhile, yet another level of standards for standards has been created by the increasing need for third-party organic certification systems to conform to International Organization for Standardization (ISO) Guide 65 processes, which set out general requirements for third-party certification systems (ISO, n.d.). As described by Mutersbaugh (2005a), third-party certification systems that do not comply with the ISO 65 template can be challenged through the WTO as constituting a trade barrier.

IFOAM's basic standards also function as a second level of standards for standards, and its accreditation is the key for international recognition in Southeast Asia. IFOAM in turn participates in a number of other international bodies that aim to harmonize international certification systems. IFOAM, the United Nations Conference on Trade and Development (UNCTAD), and the United Nations Food and Agriculture Organization (FAO) created the International Task Force on Harmonisation and Equivalence in Organic Agriculture (ITF) in 2002 to facilitate discussion and harmonization among the international organic standards (ITF n.d.). The ISEAL Alliance has developed a Code of Good Practice for Setting Social and Environmental Standards, for member organizations, which includes the Fair Trade Labeling Organization International, the Forestry Stewardship Council (FSC), and others. Although the FSC was created partly out of a recognition of the need to more effectively link social and environmental justice, the effect of these layers of standards for standards is also to continually move the templates that shape certification further and further from farmers. These moves are linked to a broader neoliberal project of working with corporations to facilitate and regulate international trade, and to commodify environmental and labor qualities for sale to consumers.

The world of organic certification, in other words, has moved a long way from the days when groups of farmers could set informal standards in accordance with their local understandings of what constituted organic. The growing significance of ISO regulations, and the development of organic standards by Codex Alimentarius for the purpose of resolving trade disputes around organic food, demonstrate how the logistical needs of a corporation have arguably reorganized certified organic farming into a disciplinary process in which farmers need to comply with global standards to obtain entry into this sector. These layered, hybrid public-private, semitransparent rules can be seen as a form of structural power that is remaking the organic agrifood industry in ways that are promarket—that is, creating a new forms of commoditization and thus opportunities for corporate accumulation.

In terms of what this means for farmers: when organic food becomes defined by certification, then producers must follow explicit technical and administrative guidelines for certification from accredited agencies, and face regular inspections to test for compliance. Participation in the organic certification economy demands investment in compliance with internationally established standards for the production of documents, inspections, and laboratory testing. Farming practices must be rendered visible and measurable. Keeping adequate records requires significant changes in farmers' practices. Other requirements (for "good agricultural practice" standards) include farm upgrades (waste treatment and disposal, toilet facilities for workers, storage structures for farm inputs); use of protective equipment; training in hygiene and safety requirements; and avoidance of using fresh manure (Nguyen et al. 2006). Certification requires inspections and documentation not only on the farm, but at each step in the commodity chain. This "audit culture" (Friedberg 2007; Hatanaka, Bain, and Busch 2005) in global food regulation reflects neoliberal reforms that some authors argue have shifted governance from state organizations to the private sector (Campbell 2005). As we described above, however, the incursions specifically related to organic production have taken place in the opposite direction, through state regulation into what began as primarily private—though not necessarily corporate—regulation.

The setting of standards, inspections, monitoring, and accreditation activities have become an industry in itself—employing workers to produce certified qualities through monitoring, and the production of documents that themselves become commodities (Mutersbaugh 2005b, 393). Programs to certify food production are in effect a labor process for the production of values that are appropriated by different actors along the commodity network. The boom in international demand for organic products has created a burgeoning field of certification bodies and inspection companies. Not all are "for profit" but they are jockeying for clients, and expanding rapidly.

Although IFOAM or Codex Alimentarius standards are intended to be flexible, and can technically be adapted when applied to local conditions, the bottom line is that international trade and corporate marketing require standardized standards. Moreover, corporate buyers interested in increasing the market competitiveness of their products can dictate the terms of what constitutes high-quality standards and impose rigid and demanding requirements above and beyond those set by national standards or IFOAM and Codex Alimentarius. Activists and concerned consumers in the global North also tend to be more concerned that Northern-based standards are properly enforced in Southern contexts—whether the product is truly organic—than that the standards are flexible and locally appropriate. Thus, there is ongoing pressure to ensure that inflexible standards are rigidly enforced in the name of maintaining consumer confidence, at least in the agribusiness-controlled distribution channels. This can occur though the structural power of retailers in making demands that producers meet set standards, either through providing evidence of certification or by meeting retailers' benchmarks of quality. Similarly, agribusiness influence over the certification process facilitates the establishment of industry standards that all producers feel the pressure to adhere to—even those producers that may not yet supply retail chains.

A number of food activists and NGOs are rejecting the industrialization and mass marketization of organic food, provoking a number of schisms in the organic or alternative farming movement. In reaction to government restrictions on the use of the term *organic*, and the corporatization of organic food production, food system activists have become more interested in promoting local food systems and direct consumer-producer links. The criticisms of activists have had an impact on IFOAM and other certified organic organizations that now recognize the need for increasing the scope for local-level flexibility to counter the exclusion of small farmers from using certified organic labels. Organizations like IFOAM are also paying more attention to social justice and labor standards, which in turn helped prompt the formation of ITF[1] and ISEAL.

These debates are particularly significant in the context of how certified organic farming has moved into the global South, where certification has become associated not just with industrial agriculture, but also with Northern influence. Because certification is especially important for long-distance supply chains, the certification of organic farming was introduced to global South producers by global North suppliers and retailers. This has reinforced the structural influence of Northern retail corporations that have closer ties with certification bodies in the North. Domestic certification systems in the global South often lack credibility (IFAD 2005), so most organic products destined for global markets require certification with international standards—or more specifically, with global North certification bodies. This leaves organic certification susceptible to an additional layer of criticism: in Thailand, for example, activists in the alternative agriculture movement argue that international organic standards were created on the basis of Northern agriculture, and that there are many obstacles to participation by small farmers in the global South. Friedberg (2007) argues that the overriding of practical, local knowledge by codes of standards ensuring safety, quality, and ethical content constitutes an imperialism that builds on other forms of imperial governance (see also Vandergeest 2007).

These debates are articulated among agrifood activists in Southeast Asia—an important site for the expansion of certified organic agriculture. The specific ways in which corporate organics has influenced agrifood systems and alternative agriculture movements depend on a series of contextual conditions, including (1) the significance of export agriculture; (2) the relative importance of supermarkets; (3) the strength of social movements, international donor agencies, and NGOs promoting ecological or alternative agriculture; and (4) the role of government. In the next section we describe emergence and context of corporate organics in Thailand, Vietnam, and Indonesia.

Organic Agriculture and Certification in Thailand, Vietnam, and Indonesia

Corporate influence in green agriculture has spread unevenly and in various forms. Below we outline the introduction and current status of certified organic production in the three countries considered here. To set a brief context for the heterogeneity between the three countries, some basic socioeconomic data is presented in table 3.1. In Southeast Asia, certified organic production comprises one part of a larger movement to promote both organic farming and so-called safe or hygienic foods, which are now common in food markets around the region. The number of farmers involved in producing "safe" foods is growing rapidly, with the support of NGOs, development agencies, and governments. Many are sold uncertified in local markets; some are certified for food safety by health departments, others by agricultural departments, development agencies, and farmer cooperatives for characteristics including "pesticide free," "good agricultural practices," or corporate quality labels. Thus, local markets—including wet markets, small shops, and supermarkets—in these three countries sell a range of "green" foods, much broader than only certified organic.

We use the term *green foods* as a catchall label for a variety of foods with value-added qualities promoted as natural or alternatives to mainstream products. These include those that indicate no chemical residues, sustainable farming, and, less often, traditional varieties. Conventional definitions of *organic* (e.g., IFAD 2005, xiii) explain it as an internationally certifiable (based on controls and traceability) farm management system that employs soil conservation measures, crop rotation, and the application of biological and manual farming methods instead of synthetic inputs. This can be contrasted with "safe" or "hygienic" foods, which are produced in accordance with maximum allowable residue limits for pesticides, nitrate and heavy metals, and microorganisms. These safe vegetable products use integrated pest management (IPM) or a lower level of synthetic inputs than conventional/intensive (green

Table 3.1
Basic socioeconomic data for Thailand, Vietnam, and Indonesia

	Indonesia	Thailand	Vietnam
Population (millions)	222.0	65.2	84.2
GDP per capita (in PPP*)	$4,321	$9,163	$3,373
FDI** inflows	$5,556	$10,756	$2,360

* Purchasing Power Parity
** Foreign Direct Investment
Source: Agriculture and Agri-Food Canada (2008)

revolution) farming, and may be domestically certified and labeled as such. While products may be labeled as clean, safe, green, or natural, the certification of such products is not necessarily clear. Finally, the products of what may be labeled as "traditional farming" increasingly obtain premiums in local and regional markets. Traditional farming may be oriented toward local consumption, and uses little or no purchased synthetic inputs; some observers refer to these products as "uncertified" organic.

Indonesia and Thailand have had a long-standing presence of strong development NGOs engaged in campaigns to increase public awareness and generate critical dialogue around the social and environmental effects of green revolution technologies. In Thailand, NGOs worked with farmers since the early 1980s on alternative agriculture and environmental conservation initiatives. The primary emphasis was on "organic agriculture" as a component of a broader search for alternatives to conventional development practice; very little attention was given to marketing "organic food."

The significance of these NGO-organized projects in Thailand for the later development of certified organic agriculture was also that the technical and managerial capacity developed in these sites made available potential "rents" (Mutersbaugh 2005a) associated with certification. In other words, the farmers and local NGOs had built up considerable capacity in technical, administrative, and marketing aspects of converting to and becoming certified in organic production. Continued NGO activities in farmer training and marketing up to the present arguably create a kind of subsidy that can be appropriated by more market-oriented certification actors.

As noted above, in Southeast Asia, "alternative agriculture" practices emerged in response to the chemical-intensive green revolution agriculture and were first oriented to subsistence production or local markets. Later, buyers from Europe seeking to buy organic crops approached farmers in areas where NGOs had already been working at promoting less chemical-intensive production. As organics became mainstreamed in the global North, this demand gradually led to initiatives to certify production, with donor and NGO assistance.

Market-Driven Organics

There are elements within the alternative agriculture movement in Southeast Asia that are very interested in "market access" for organic products. For these actors, value added is important to getting that access, and discipline (compliance with standards) is the means by which value added is attained. As shown below, certification was introduced by corporate buyers from abroad, and then facilitated by the expansion of supermarkets in Southeast Asia and the growing domestic market for safe and organic foods. In other words, the involvement of corporations made certification necessary. The implications of this led to greater debate (and dissension) within the broader alternative agriculture movement.

The story of what is likely the first case of certified organic farming in Thailand illustrates its corporate association.[2] In 1989, Capital Rice, part of the Thai conglomerate STC and one of Thailand's major rice exporters, was approached by an Italian buyer (Riseria Monferrato S.R.L.) about the possibility of producing organic rice. Capital Rice in turn contracted the Chiangmai-based Chaiwiwat Agro-Industrial Company to handle the production of the organic rice, with technical information provided by the Italian buyer regarding the creation of organic and associated documentation as required by the EU and IFOAM. The group identified sites in mountainous valleys in Chiangrai and Payao provinces that they understood to be relatively untouched by agrichemicals. The government's Department of Agriculture also participated in the project, conducting extension work with farmers, and then developing their own government standards for organic rice under what would later become the government's Organic Thailand label (recently changed to "Organic Q"). The requirements built into the international standards regarding research and conversion periods meant that it was not until 1995 that Capital Rice started to sell organic rice under its Great Harvest label, mostly distributed through the Italian buyer throughout Europe, and certified by Bioagricert. Capital Rice has today branched out into other organic food products, and Bioagricert remains one of the major certifiers in Thailand, often working with government departments to expand organic farming.

Expanding market access for organic foods is often presented as a "solution" to farmers and agribusinesses trying to cope with global market challenges. Thailand has been leading the region in the production and certification of "green" foods, which are marketed both for export and domestically. The international demand for these value-added products is spilling over onto domestic consumers in Thailand, to the point where the market of certified organic foods within Thailand bypassed the export market as of 2006 (Ratanawaraha et al. 2007). Visits to supermarkets show that certified organic is only the tip of the iceberg with respect to a wide range of green food products certified as "natural" or "safe." The strength of the NGO movement in Thailand accounts for the establishment of Organic Agriculture Certification Thailand (ACT) as a member of the Alternative Agriculture Network (AAN). ACT is Southeast Asia's only IFOAM and ISO 65 accredited certification organization, and offers certification for EU, US, and Japanese national standards. It recently expanded its operations to establish itself as a regional services certification and inspection body, active through much of Southeast Asia. ACT can also inspect for other certification programs, and trains other inspectors.

Certified organic agriculture in Vietnam began in the early 2000s, and constitutes a mainly export-oriented enclave, subsidized by development donors, with some support from a government keen to promote exports. Vietnam produces certified organic vegetables, herbs and spices (cinnamon, star anise, ginger), tea, fruit, shrimp, catfish (*pangasius*), cashew nuts, and essential oils. As yet there is a very limited domestic market for organic produce, although national organic standards were introduced in 2007. A strong and growing domestic demand does exist for "safe" food, particularly vegetables and meat. Safe vegetables are produced by commercial growers, near urban areas, for urban markets. The organization and certification of these foods in the late 1990s followed from a string of food safety crises, most prominently associated with pesticide residues. In the larger cities of Vietnam, a range of safe produce and poultry is increasingly available in supermarkets and some wet markets and small shops (Ho and Dao 2006; Moustier et al. 2006). The production of "safe" vegetables in Vietnam has been clearly market-driven rather than for subsistence or as part of a food sovereignty movement.

The push for "safe" or "clean" vegetables and meats for the domestic market has also received impetus from the Vietnamese state, in collaboration with local-level farmers' cooperatives[3] and foreign and domestically owned supermarkets[4] (Phan 2006). Vietnamese-owned Saigon Coop Mart has a reputation for promoting high-quality Vietnamese goods, and was the first supermarket to collaborate with an agricultural cooperative and the Department of Agriculture and Rural Development to market safe vegetables in Ho Chi Minh City in the late 1990s. Safe vegetables are now carried by foreign supermarkets in Vietnam, including the French-owned Big C, and the German-owned Metro Cash and Carry.

Certified organic agriculture began in Indonesia in the early 2000s. Indonesia's organic sector is divided between an export-oriented production sector that is located mainly off Java, and a sector of food crops on Java that targets the domestic market. Even though organic production is limited and there are no reliable statistics for the amount of production actually taking place, general and specific industry media coverage concur that the market for organics within Indonesia and for Indonesian exports has been increasing and is expected to continue doing so. One representative of BIOCERT provided a rough estimate of the value of organic production in Indonesia as US$250 million.[5]

In the Indonesian case, for producers who have long engaged in export-oriented agriculture, the incentive for organic certification may come from two factors: retaining existing market share, particularly if the export markets are becoming more stringent about standards; and the reverse situation: to preempt the above problem of hanging on to their existing market share by offering product diversification, both conventional and organic. Thus, there are two kinds of niches in the same market. In some cases, as with Indonesian export crops, it is not so much a move toward exporting organics as it is offering organics to already existing foreign buyers, given that some producers are already export oriented.

Organic production in Indonesia is split into two streams. There is the export-oriented sector, mostly on islands other than Java, of products certified by representatives of international bodies and meeting international standards. These export crops make up the bulk of the certified organic products in Indonesia. These include cashews, coffee, spices, and shrimp. There is also a domestic market–oriented food crop sector on Java. The latter is composed of rice and vegetable products, which are either uncertified but labeled "organic," or certified by Indonesian companies or agencies, according to national standards or private industry standards (set by individual certification companies). In contrast to Thailand and Vietnam, products for the domestic market in Indonesia are not uniformly labeled "safe" or "hygienic." Producers use a variety of terms, including "healthy," occasionally "chemical free," and most commonly "pesticide free." Generally, the products that are sold domestically in Indonesia are advertised as grown with "alternative" or sustainable agricultural methods. The majority of these products remain uncertified because most Indonesian producers and retailers view domestic consumers as less demanding with regard to international standards than consumers in Japan, North America, or Europe However, there are signs that this may be changing because domestic (nationally based) certification is developing quickly. As a first step, in 2002, the National Standardization Agency issued Indonesian National Standards for organic food systems. Thus, although certification in Indonesia only occurs with a small percentage of agricultural crops, both the export and domestic market production sectors appear to be on the increase, particularly as a result of growing buyer demands from retailers.[6]

National data in Thailand, Vietnam, and Indonesia on the volume and value of overall organic production, on organic exports, and on the scale of farm operations tends to be unsystematic and lagging behind the rapid increases that we observe in both export and domestic sale of organic food. Table 3.2 shows official figures on the extent of organic land area including "in conversion" land area for selected countries in Asia. However, this table arguably understates the rapid growth of certified organic in these countries by a considerable amount. Koen den Braber, an advisor with the Agricultural Development Denmark Asia (ADDA) organic agriculture project and formerly involved with Hanoi Organics and Ecolink Tea Company, estimated that there are probably an additional 6,000–7,000 hectares of land under organic management in Vietnam that are not accounted for in the IFOAM figures (Den Braber and Hoang 2007). Moreover, a brief search in the online business section of Thailand's English-language daily newspaper, *The Nation*, for example, turns up the following indicators of how organic is rapidly becoming a normal form of production for corporations: "The country's largest sugar refiner, Wangkanai Group, has decided to convert all of its 20,000 contract farmers to organic by 2010, beginning with a program to train 800 'leaders' among these farmers" (*The Nation* 2006). In another instance: "A major organic exporter, Swift Co., now wants to capture the domestic market to reduce risks inherent in overseas sales. Its brand will be sold in supermarkets and by direct sales; the company plans to have more than 100 centres in the Bangkok area to handle direct sales. More than 10,000 family farms are employed as contract growers" (*The Nation* 2007).

The above news item, and data cited earlier showing that domestic markets for certified organic in Thailand are now greater than exports, underscore the shift toward domestic markets that is driven by the very

Table 3.2
Area under organic management (fully converted and "in conversion") and organic farms, for selected countries in Asia

Country	Year	Organic land area (hectares)	Share of organic land (percent)	Number of organic farms
China	2005	2,300,000	0.41	1,600
India	2005	150,790	0.08	5,147
Indonesia	2005	17,783	0.04	15,473
Japan	2005	8,109	0.16	4,636
Korea	2005	38,282	2.01	5,447
Malaysia	2005	962	0.01	40
Philippines	2004	14,134	0.12	34,990
Thailand	2005	21,701	0.12	2,498
Vietnam*	2001	6,475	0.07	1,022

* The fact that the 2007 edition of this report has a 2001 figure listed for Vietnam suggests that the government is not collecting data on this. Den Braber and Hoang (2007) estimate that the actual area of organic land is probably 6,000 to 7,000 hectares higher.
Source: Ong Kung Wai (2007: 111–112).

high premiums available for certified foods. Our own supermarket surveys in the three countries confirm Roitner-Schobesberger et al.'s (2008, 117) statement that most organic vegetables obtain premiums of 100 to 170 percent above nonlabeled produce, and 50 percent compared to vegetables labeled hygienic or safe (under government certification). A survey of supermarkets and wet markets in Chiangmai, Thailand, during 2005 showed popular vegetables like Chinese kale and morning glory carrying premiums of up to 400 percent for certified organic; premiums for rice were lower but certified organic nevertheless still obtained a premium of close to 100 percent above equivalent grade and variety uncertified rice. In Vietnam, safe vegetables typically garnered a premium of at least 20 percent higher than regular vegetables.

In 2003, there were 93 certification bodies in Asia, as listed in the Organic Certification Directory. In 2005, this number had risen to 117, 104 of which were located in China, India, and Japan (Rundgren 2007). Certification bodies can be governmental bodies, private entities, or foreign entities. Many Asian countries still do not have local service certification providers. There is a growing push to have in-country professionals for inspecting and certifying, to overcome the high costs of certification. But in the meantime, the majority of certified organic production is carried out by foreign certification bodies such as the Swiss-based Institute for Marketecology (IMO)[7], the French-based Ecocert, or the Dutch-based Control International, rather than by national agencies. These bodies can certify products according to a range of international standards, depending on the final market for the goods. Some of the common governmental regulatory standards for organic production are the USDA National Organic Program, the Japanese Agricultural Standards, and the EEC Regulation No. 2092/91 of the EU. A range of private organic standards are also common, such as Naturland, Bio-Suisse, and Aquaculture Certification Council (see table 3.3).

Certified organic farming in Southeast Asia involves a wide range of farmers and relationships, including small farmers under group certification; corporate farms similar to those emphasized in some North American literature (Guthman 2004); smaller contract farmers; and farmers enlisted through government projects. The presence of larger corporate farms is indicated by the lists of farms certified by a major EU-based certification company in Thailand: it includes a 600-hectare oil palm farm, a 600-hectare sugarcane farm, and a 270-hectare farm certified for cassava. At the other end of the spectrum are smallholders under contracts that specify organic or similar arrangements. Rice, one

Table 3.3
Some common certification bodies and organic standards used in Thailand, Vietnam, and Indonesia, with countries of origin

International certification bodies	Food quality standards
ACT (Organic Agriculture Certification Thailand)	*Foreign governmental standards* EEC Regulation No. 2092/91 (EU)
IMO (Institute for Marketecology) (Switzerland)	USDA National Organic Program (NOP) (USA)
Ecocert (France)	Japanese Agricultural Standards (JAS) (Japan)
Control International (formerly SKAL International) (Netherlands)	
Bioagricert (Italy)	*Foreign private standards*
ICEA (Instituto per la Certificazione Etica ed Ambientale) (Italy)	Naturland (Germany) Bio-Suisse (Switzerland)
The Soil Association (UK)	KRAV (Sweden)

of Thailand's major certified organic products, most often appears in certified lists through group certifications; thus the same EU-based certification company lists over 1,000 hectares of paddy certified through an agricultural cooperative. ACT's list of certified operators as of March 2005 (see ACT n.d.) includes five grower groups with over 100 members each, and more groups with smaller numbers of members, with rice the most common product. In addition, as the above reports from the Thai media indicate, a significant portion of the market-oriented organic farming is comprised of smallholders who do contract farming through corporate buyers. Finally, many development programs run by the government and by international development agencies (the Danish agency, DANIDA, among others) are now adopting certified organic farming.

The rise of certified organic food production among small farmers and small and medium-sized enterprises in Southeast Asia has been facilitated by a range of funding and technical support programs, provided by agencies such as the Swiss Import Promotion Programme, the German development agency GTZ, DANIDA, the USAID-supported Vietnam Competitiveness Initiative, and AusAID. Some companies use a smallholder group certification mechanism and internal control system (ICS) to enable certification at a lower cost for small-scale producers in developing countries.[8] One example of this is the ICS established by Ecolink with organic tea farmers in Thai Nguyen province in Vietnam.

Another recent development in green certification is the establishment of participatory and pro-poor food quality guarantee systems. Initiatives such as these are often supported by international development donors, thereby subsidizing production and potential appropriation by corporate buyers. The principle behind this is to allow poor producers to earn higher profits from supermarket-driven and other value chains by enabling them to demonstrate the quality of their food products through certification at an affordable cost. A formal certification system provided by departments of agriculture at the provincial level may not meet the level of reliability demanded by urban consumers concerned about food safety. Thus, one collaboration in Vietnam between the Asian Development Bank's Making Markets Work Better for the Poor (M4P) program, CIRAD (Centre de Coopération Internationale en Recherche Agronomique pour le Développement), and an NGO, ADDA, is seeking to develop a pro-poor food quality guarantee system that is geared to small-scale producers, targeted at local consumers, and developed in a participatory manner (M4P et al. 2006). This is based on IFOAM's experience with Participatory Guarantee Systems. However, the same argument could be leveled here as made earlier, that such certification programs open the door to corporate influence in the organics sector, albeit through engaging smaller-scale actors supplying these retailers and export companies.

Specificities of Organic-Sector Development in Southeast Asia

The production of certified organic food and agriculture has grown substantially over the past ten years in all three countries considered here. The influence of corporate agrifood actors has seen a parallel growth, primarily through structural and discursive forms of power, which have reinforced one another. At the same time, there is considerable variation across these countries, and among regions within them, in the rate of growth; the relative orientation to domestic versus export markets; and the forms that organic farming takes. Although the reasons for these differences are complex, we discuss three conditions that account for some of this variation. These are the (1) the growth of the corporate retail sector, linked to the growth of urban middle classes; (2) government support; and (3) the way corporate influence and organic certification have been contested. The constellation of actors shaping the certification of green foods is illustrated in figure 3.1.

Corporate Retailing, Supermarkets, and the Middle Class

Southeast Asia is a region of rapid urbanization and a growing middle class. The increased employment opportunities for women and growing proportion of nuclear families mean that there are greater opportunity costs on women's time. These factors make ready-to-eat and semi-prepared food items more appealing than preparing meals from scratch. Urban consumers have begun to rely less on the traditional "wet" markets and instead on corporate retailers (supermarkets, hypermarkets, convenience store chains, and so on) stocked with large volumes of ready-to-eat, snack, and highly processed foods. These trends are reinforced by the changing lifestyles and diets, and the increased demand for out-of-season and imported foods. The growing use of credit cards, cars, and refrigerators further propels the shift to corporate retailers as opposed to "mom-and-pop" shops and traditional markets (Phan 2006). To the extent that domestic populations in Southeast Asia increasingly buy their

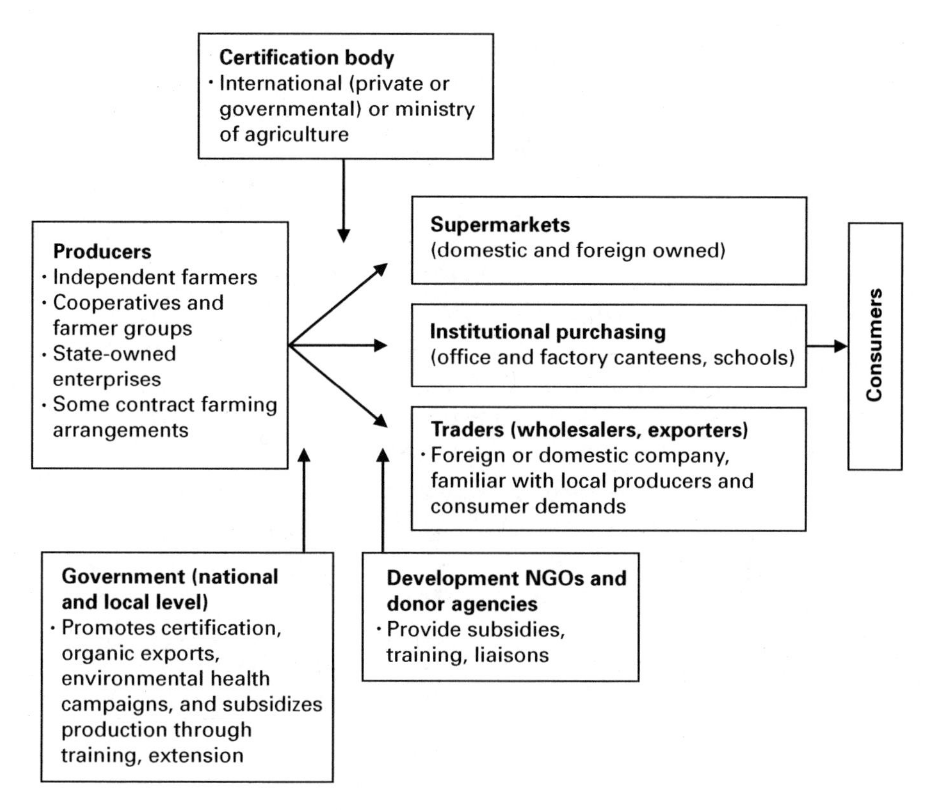

Figure 3.1
Actors involved in the making of certified organic and "green" foods in Southeast Asia

goods in corporate retail outlets, this can be seen as evidence of the discursive power of corporations, because the latter are able to pitch themselves as providers of higher quality and safer food.

Corporate retailers are perceived as the logical retail outlets for higher-value products such as organic food. Organic foods are sold on a construction of "quality" characteristics—through the discursive power of their advertising—and corporate retailers, with their concentrated position in the retail market, possess the buying power to reinforce these quality standards. In Vietnam, 60 percent the country's eighty-two million people are under the age of thirty (Agriculture and Agri-Food Canada 2004), and it is this group—particularly young urban residents with sufficient disposable income—that is most receptive to new food trends and higher-value products such as organic foods. Moreover, a recent study found that 88 percent of surveyed Hanoi residents considered vegetables to be a health hazard due to the increased use of agrichemical inputs (Figuié et al. 2004). This is indicative of the growing consumer concern about the safety of the food supply, a further impetus for consumers to purchase safe or organic foods at high premiums.

Prior to the late 1990s, expansion of the supermarket/hypermarket sector was secured through Asian investment, both domestic and Japanese (Commonwealth of Australia 2002). Since the late 1990s, this sector has experienced accelerated expansion, increasingly driven by foreign chains, although domestic chains continue to operate and in some cases expand as well. Major transnational investors include Carrefour (France), Casino Guichard-Perrachon (France), Royal Ahold (Netherlands), SHV Holdings Makro (Netherlands), Delhaize Group (Belgium), Tesco (UK), and Metro (Germany) (Commonwealth of Australia 2002; Wiboonpongse and Sriboonchitta, 2007). There are also some important locally owned supermarkets, including Central Group and Charoen Pokphand (in Thailand), Maximark, Coop-mart, and Intimex (in Vietnam), and Hero and Matahari (in Indonesia).

It is striking the degree to which European-based supermarkets dominate in Southeast Asia. This is significant because European retailers are also leaders in introducing private standards, and Europe is a major market for organic products (Sahota 2007; Barrett et al. 2002). European regulation on mergers and acquisitions in the early 1990s inhibited further corporate concentration of retail holdings in the EU. The solution for companies like Tesco, Metro, and Carrefour, among others, was to expand overseas. Following the 1997 Asian financial crisis, the structural power of corporations was strengthened, particularly in terms of the European corporate presence in agrifood commodity chains in Southeast Asia. The 1997 crisis made Asian companies cheap for foreign direct investment, and many of these European corporations then bought up shares of struggling Southeast Asian companies as a result. This was also made politically possible because the crisis gave further impetus to liberalization and decreased protectionism, which in some cases was part of the conditionality of International Monetary Fund loans during the crisis period.

The corporate dominance of the retail sector, particularly supermarket chains, has been key in strengthening the structural power of corporate players in the green foods sector in Southeast Asia, in terms of demanding uniform certified products, and to some extent in shaping private organic certification initiatives.

Government Roles

The governments of Thailand, Vietnam, and Indonesia are engaged in a number of activities regarding organic agriculture, which implicitly facilitate corporations' influence in the organic sector. First, they have promoted green certification, including aquaculture, as part of a broader export strategy promoting high value-added agrifood exports. This promotion involves a range of measures including farmer training, using development funds to pay private certification companies to certify farmers, and assistance with international marketing. The facilitation of certified organic exports includes the creation of national standards and accreditation procedures for certification bodies. All three countries have developed national organic standards and procedures for certification, even if few products are currently certified for domestic markets in Vietnam and Indonesia due to limited consumer demand in these countries. Indonesia's and Thailand's national organic standards are drawn from the Codex Alimentarius (established in 1991 as a joint FAO/WHO standards program) and the IFOAM. In both cases, government standards are being revised to allow for adapting the standards to local conditions.[9]

In Thailand, government agencies have constituted themselves as certifying bodies—a role that is contested by activists who accuse the government of a conflict of interest. So far only the government of Thailand has created its own brand/label. Although not recognized internationally by IFOAM, products with this brand are sold in local supermarkets and purchased by crown corporations such as Thai Airways. The Thai government entered the field of organic certification in 2000, with the establishment of the Organic Crop Institute, which later shifted to the National Bureau for Food and Agriculture Commodity Standards, established in 2002 (Ratanawaraha et al. 2007). In 2005, the government created the National Agenda for Organic Agriculture, which involved multiple ministries and departments, with a goal of reducing the volume of imported agrichemicals through converting 850,000 farmers to organic agriculture within five years. This agenda has been given an additional push by the government's commitment to the King's Sufficiency Economy initiative. The government promotes ecological agriculture through major extension and training programs, a government certification program for both farmers and retailers/wet markets, a marketing campaign, export promotion, and by using its influence among key locally owned or crown corporations to push the expansion of organic and green production/retailing. Particularly significant is an expanding program to train farmers in organic farming practices. According to interviews conducted in June 2007, the Ministry of Agriculture was completing a program that funded 40 organic farmers to train 40,000 more farmers; the plan was to expand this program the following year to enlist 300 organic farmers to train 150,000 more farmers. The ministry also has a key role in publicizing the importance of ecological agriculture and safe food.

The Indonesian government undertook a program called "Go Organic 2010" aimed at raising national awareness of the benefits of organic farming. The Department of Agriculture is promoting the growth of organic agriculture through this initiative as a form of "eco-agribusiness" targeting primarily export but also domestic markets.

The role for government in promoting safe agricultural products in Vietnam includes training in IPM and testing and certification of safe foods. Testing is periodically carried out for heavy metal content and pesticide residues (Ho and Dao 2006). While safe foods are being promoted by the government, organic certification appears to be mainly driven by the private sector in liaison with some NGOs (such as ADDA), local government authorities, and the national Farmers' Association, which supervise the production and packaging in conjunction with the Department of Plant Protection.

Earlier in this chapter we noted the increasing roles of government and intergovernmental organizations in certification. This trend reinforces the removal of farmers or even national-level farmer organizations from the role of setting standards, when it is these groups who generally understand local and national production conditions best. Moreover, government involvement in establishing green certification opens a path for corporate retailers to procure and sell certified green foods at premium prices.

Contesting Corporate Influence in Organic Production

Local as well as transnational NGOs and international donor agencies have been divided over the expansion of market-oriented organic production and the role of corporate players in this process. A discursive struggle between corporations and the alternative agriculture movement has emerged, with the former attempting to capture the discourse around safe and organic foods in order to increase market share. This struggle has led to differing versions of what is "sustainable" in terms of organic production. Further, the involvement of some NGOs in organic certification schemes has effectively subsidized corporations by lowering certification costs, thus further entrenching their dominance in the market.

There are a number of organizations that see the future of green agriculture through engagement with corporations and international markets. In Thailand, Green Net (and its influential director, Vitoon Ruenglertpanyakul) supports the move to export-oriented organic agriculture, as does ACT. Fair trade networks also promote international marketing of their organic products, and sometimes sell through large retail chains. In Vietnam, ADDA is an NGO that has provided substantial support for the development and marketing of organic products. Many of the organic products in Vietnam are farmed by small producers and coordinated by small domestic companies, sometimes in liaison with an NGO or international donor agency that facilitates the contact with international buyers (e.g., a supermarket chain in Europe).

A contrasting position of some NGOs vis-à-vis corporate engagement is to reject the market orientation of organic agriculture and the propagation of international standards for large-scale and export-oriented organic production as potentially undermining the rural ways of life that they seek to support. This position can be found among both local and transnational NGOs. Transnational NGOs that are critical of corporate engagement include Via Campesina and Focus on the Global South. Within Thailand, some members of the AAN are critical of this shift, including, prominently, the Northern Alternative Agriculture Network. Most of these organizations promote the importance of self-sufficiency, and diversified production for local markets. In some cases, they also facilitate consumer-producer linkages, and certification based on local standards.

Perhaps more so than in the global North, NGOs have played key roles in the development of alternative agriculture practices, which led into what is now known as organic farming. Yet ironically, when farmers who have been working with NGOs turn to selling certified green products to corporate buyers, corporate rents are being created by the work of NGOs and farmer groups. Thus, like the role of the state, the strong presence of local organizations and international development agencies in promoting greater roles for smallholders in certified organic production is serving to bring down (i.e., subsidize) the costs of certification.

Many of the key debates within the alternative agriculture movement now turn around the question of international certification, arguably an outcome of corporate appropriation of organic agriculture and food. Indeed, even the initial anti–green revolution movement, to the extent to which it is anticorporate, is shaped by corporations (e.g., Delforge 2004). However, even within this critical discourse, there are differing views as to how green production can address the problem of farmer vulnerability vis-à-vis other social agents (commodity buyers, corporate retailers) in the commodity chain. Some argue that rather than empowering farmers (as envisioned by alternative agriculture advocates), green production subjects farmers to greater external control through complying with strict standards of the organic food industry. Thiers's (2005, 11) insights regarding the hierarchical system of research and extension in China are echoed to a large degree in Vietnam: "There is only limited evidence that farmers themselves have any influence on the research agenda. . . . In the culture of Chinese agricultural development . . . the top-down approach is deeply ingrained and local governments exercise considerable authority. This issue is further complicated by the emphasis on quality control and enforcement of standards inherent in organic food certification and marketing." In other words, "The strict demands of the organic food industry may eclipse the ecological principles of organic farming, particularly where the need for 'quality control' can be used as a justification for local authoritarianism" (Thiers 2005, 11).

The presence of other social agents in the making of green agriculture can influence the struggle over corporate control in these three countries. The role of the state is important in all three countries, but Vietnam lacks the kind of domestic NGOs and farmer groups present in Thailand and Indonesia, which challenge and lobby the state over the direction of agricultural policies. These groups can serve as a counterweight or at least exert a moderating effect to corporate influence. In Vietnam, in contrast, although some international NGOs are quite vocal, overall there is a lack of grassroots organizing. However, strong pro-corporate sentiments are still present among certain state institutions in Thailand and Indonesia, as witnessed in both countries' policies toward the promotion of export-oriented agribusiness development. The important distinction between these cases and that of Vietnam is that the alternative agriculture movements in Thailand and Indonesia are characterized by a wider spectrum of actors with differing agendas, resulting highly divisive debates over the degree of corporate control over the green foods sector. In Vietnam, the state plays a singularly more prominent role since there are few actors capable of challenging the state on agrifood policy on the same basis as in Thailand and Indonesia. For green agriculture, this has resulted in an approach that is primarily driven by the state agenda of improving food safety and health, along with continued access to profitable export markets.

Conclusion

Corporate influence has expanded both in structures of governance of agriculture and food and in commodity chains, from inputs and production to retailing (Fold and Pritchard 2005; Marsden, Harrison, and Flynn 2000; Higgins and Lawrence 2005). This corporate influence has now expanded into the production and retailing of organic and other "green" foods in international markets. The setting of norms for consumer preferences is evidence of the discursive power of corporations. A related facet of this power concerns the requirements for certification of green foods.[10] Thus, what once constituted "alternative" production approaches and retail networks are increasingly falling under the dominion of private, and often corporate, agrifood chains. In the process of "mainstreaming" corporate organics, less of the value added is fed back into local economies, and farmers have less autonomy over the process—although some counterefforts have challenged these broader trends.

This chapter has traced the shift in "alternative" agriculture and "green" food products in Southeast Asia from its original domestic orientation—as a reaction to and critique of "green revolution" approaches—to becoming the subject of growing international demand and strict certification standards. Subsequently, and to some extent in response to this trend, there has also been a reorientation of "safe" if not organic food products for the domestic market to meet the demands of the growing middle classes for healthy and safe—and, to an extent, locally branded—foods. In the backdrop of these shifts is the role of corporations, first in promoting agrichemical inputs for green revolution agriculture, which in turn was an impetus for organic farming, as a countermovement. Later, corporations—and supermarkets in particular—became important buyers of organic products, and promoted their certification. These trends reflect neoliberal objectives of trade promotion and commodification of green qualities for consumer purchase—both important elements in enhancing the structural power of corporations.

Although the boom in organics globally may be attributed most recently to the spreading of certification programs, we argue that the origins of this shift away from chemical-intensive production lie in sustainable local-development initiatives at the community scale, with collaboration between farmers, activists (social movements), and NGOs (Allen and Kovach 2000; Raynolds 2004). The corporatization of organic foods spurred an anticorporate reaction among farmers' groups and consumers, who in turn began to promote locally sourced and domestically certified goods as an alternative to corporate-driven supply chains. This reaction has also raised debates over whether the alternative agriculture movement should promote the export of certified organic products, and over who should participate in setting organic and related standards.

Our analysis looked at these dynamics in a Southeast Asian context, and specifically identified how the activities of noncorporate actors can reinforce—or sometimes counter—the power and influence of corporate interests. We showed that the governments of Thailand, Vietnam, and Indonesia—together with some development and donor agencies—have played an important role in facilitating the expansion of organic certification, which has been paralleled by an increasing role for corporate retailers in purchasing organic foods, for both international and domestic markets. NGOs and farmer organizations have been divided over this issue of international certification and expansion of market access for organic foods. While some may be content that the end product is an organic product, critics point out that the discursive power of corporations has implications for sustainability insofar as local knowledge and practices become increasingly marginalized and subservient to both market forces and demands from corporate buyers. This is in contrast to a system of organic production oriented to local markets and based on direct producer-consumer linkages and relations of trust, in which the cost of social and environmental sustainability is internalized in the production process (Clark 2007).

In sum, the current promotion of certification by global South governments in particular, but also among some NGOs, can only be understood in relation to the structural power of the corporate agrifood industry. Without the demand for certified organic from abroad, and its subsequent adoption by supermarkets in Southeast Asia as well, it is not likely that the Thai—and to a lesser extent, the Vietnamese and Indonesian—governments would be involved in promoting organic certification.

Indeed, the emerging centrality of certified organic farming for sustainable, alternative, and locally oriented agriculture can arguably be traced to the power of corporate buyers. Further, corporations have utilized a discursive strategy in order to increase their market share, by shaping public perceptions of their role as providers of organic and safe foods. This has sparked a discursive struggle between corporations and anti-corporate NGOs and social movements regarding the "sustainability" of corporate-based organic agriculture. At the same time, corporations have to some extent benefited from the involvement of some NGOs and government actors in the certification process, which has effectively lowered certification costs.

Acknowledgments

Thanks to Erin Nelson and Jennifer Clapp for constructive feedback on an earlier version of this chapter. Funding for some of the fieldwork conducted for this research was provided by the Challenges of the Agrarian Transition in Southeast Asia project, supported by the Social Sciences and Humanities Research Council of Canada.

Notes

1. "A plethora of certification requirements and regulations are considered to be a major obstacle for continuous and rapid development of the organic sector, especially in developing countries. Sharing great concern in this regard, IFOAM, FAO and UNCTAD decided to join forces and agreed that harmonization, mutual recognition and equivalency in the organic sector offer the only viable solution to overcome the problems indicated above" (ITF, n.d.)

2. Based on interviews on June 1, 2007, and on Vandenberghe and Sarakosas 1997.

3. Until the 1980s, Vietnam's agricultural production was largely organized into agricultural collectives, in which land and labor were pooled. These were subsequently disbanded to allow for the emergence of a market-oriented economy and allocation of land directly to households. A large number of voluntary cooperatives have since been formed, but mainly for coordinating agricultural services and inputs, while production is managed by each household, which has been issued a long-term land use-right certificate (Scott 2008).

4. The number of supermarkets in Vietnam grew from 2 in 1999 to over 100 in 2006. These food outlets account for approximately 10 percent of food sales (Agriculture and Agri-Food Canada 2008).

5. Interview with BIOCERT representative, June 14, 2007.

6. Specialty chain stores such as Healthy Time in Jakarta advertise on their premises the requirement for all their produce to be certified, though they do not specify by which standards.

7. As of 2007, IMO Vietnam was engaged in consulting, training, surveys, and inspection, but certification was performed by IMO Switzerland.

8. Farmer groups in some parts of Thailand have rejected the model of smallholder group certification for perpetuating a culture of mistrust among farmer groups.

9. International standards are usually phrased broadly. A variety of agencies—government, national certification bodies, and standards organizations—as well as local inspectors for these agencies, decide how they are to be interpreted for the following two purposes. First, international standards can be the basis for the construction of newly established national standards. Second, when local producers apply for certification by international standards, in the course of conducting inspections, the inspectors will have to ensure that locally specific practices are compliant with the broadly framed intentions and practices set out in the international guidelines.

10. Although IFOAM now emphasizes flexibility, the daily work of these certifying bodies involves monitoring farmers to ensure they meet standards specified by importers and importing companies, which are relatively inflexible.

References

ACT (Organic Agriculture Certification Thailand). n.d. List of ACT Certified Operators. http://eng.actorganic-cert.or.th/list_oper.html.

Agriculture and Agri-Food Canada. 2004. Vietnam Agri-Food Country Profile Statistical Overview. http://atn-riae.agr.ca/asean/3834_e.htm.

Agriculture and Agri-Food Canada. 2008. *Agri-Food Regional Profile: ASEAN*. Ottawa: Agriculture and Agri-Food Canada.

Allen, Patricia, and Martin Kovach. 2000. The Capitalist Composition of Organic: The Potential of Markets in Fulfilling the Promise of Organic Agriculture. *Agriculture and Human Values* 17 (3): 221–232.

Barrett, H. R., A. W. Browne, P. J. C. Harris, and K. Cadoret. 2002. Organic Certification and the UK Market: Organic Imports form Developing Countries. *Food Policy* 27 (4): 301–318.

Buck, Daniel, Christina Getz, and Julie Guthman. 1997. From Farm to Table: The Organic Vegetable Commodity Chain of Northern California. *Sociologia Ruralis* 37 (1): 3–20.

Campbell, Hugh. 2005. The Rise and Rise of EurepGAP: European (Re)invention of Colonial Food Relations? *International Journal of Sociology of Food and Agriculture* 13 (2): 1–19.

Campbell, Hugh, and Annie Stuart. 2005. Disciplining the Organic Commodity. In Vaughan Higgins and Geoffrey Lawrence, eds., *Agricultural Governance: Globalization and the New Politics of Regulation*, 84–97. London: Routledge.

Clark, Lisa. 2007. Business as Usual? Corporatization and the Changing Role of Social Reproduction. In the Organic Agro-Food Sector. *Studies in Political Economy* 80: 55–74.

Commonwealth of Australia. 2002. *Subsistence to Supermarket II: Agrifood Globalization and Asia. Volume 1: Agrifood Multinationals in Australia*. Canberra: Department of Foreign Affairs and Trade.

Delforge, Isabelle. 2004. *Thailand: From the Kitchen of the World to Food Sovereignty*. Bangkok: Focus on the Global South. www.focusweb.org.

Den Braber, Koen, and Hoang Thi Thu Huong. 2007. *Feasibility Study for Organic Bitter Tea Production and Marketing in Cao Bang Province*. Cao Bang, Vietnam: Helvetas Vietnam. December.

Figuié, M., N. Bricas, Vu Pham Nguyen Thanh, and Nguyen Duc Truyen. 2004. Hanoi Consumers' Point of View Regarding Food Safety Risks: An Approach in Terms of Social Representation. Presentation to the XI World Congress of Rural Sociology, Trondheim, Norway, July 25–30.

Fold, Niels, and Bill Pritchard, eds. 2005. *Cross-Continental Agro-Food Chains Structures: Actors and Dynamics in the Global Food System*. London: Routledge.

Friedberg, Susanne. 2007. Supermarkets and Imperial Knowledge. *Cultural Geographies* 14 (3): 321–342.

Fulponi, Linda. 2007. The Globalization of Private Standards and the Agri-food System. In J.F.M. Swinnen, ed., *Global Supply Chains, Standards and the Poor*, 5–18. Oxfordshire: CABI.

Global Industry Analysts. 2007. *Organic Foods & Beverages: A Global Strategic Business Report*. San Jose, CA: Global Industry Analysts.

Guthman, Julie. 2004. *Agrarian Dreams: The Paradox of Organic Farming in California*. Berkeley: University of California Press.

Hatanaka, Maki, Carmen Bain, and Lawrence Busch. 2005. Third-Party Certification in the Global Agrifood System. *Food Policy* 30 (3): 354–369.

Higgins, Vaughan, and Geoffrey Lawrence, eds. 2005. *Agricultural Governance: Globalization and the New Politics of Regulation*. London: Routledge.

Ho, Thanh Son, and Dao The Anh. 2006. *Analysis of Safe Vegetables Value Chain in Hanoi Province*. Hanoi: Report for Metro Cash and Carry Vietnam, GTZ, and Vietnam Ministry of Trade.

International Fund for Agriculture and Development (IFAD). 2005. *Organic Agriculture and Poverty Reduction in Asia: China and India Focus*. Report No. 1664. Rome: IFAD.

International Organization for Standardization (ISO). n.d. http://www.iso.org/iso/iso_catalogue/catalogue_tc/catalogue_detail.htm?csnumber=26796.

International Task Force on Harmonisation and Equivalence in Organic Agriculture (ITF). n.d. http://www.unctad.org/trade_env/projectITF.asp.

Johannsen, Julia, Birgit Wilhelm, and Florian Schone. 2005. *Organic Farming: A Contribution to Sustainable Poverty Alleviation in Developing Countries?* Bonn: German NGO Forum on Environment and Development.

Lang, Tim, and Michael Heasman. 2004. *Food Wars: The Global Battle for Mouths, Minds and Markets*. London: Earthscan.

Making Markets Work Better for the Poor (M4P)–Asian Development Bank, MALICA-CIRAD (Centre de coopération internationale en recherche agronomique pour le développement), and ADDA (Agricultural Development Denmark Asia). 2006. Setting a Participatory and Pro-Poor Food Quality Guarantee System: Draft Proposal. March 21. http://markts4poor.org/?name=currentwork&op=viewDetailNews&id=1121.

Marsden, Terry, Michelle Harrison, and Andrew Flynn. 2000. *Consuming Interests: The Social Provision of Foods*. London: UCL Press.

Moustier, Paule, Muriel Figuié, Nguyen Thi Tan Loc, and Ho Thanh Son. 2006. The Role of Coordination in the Safe and Organic Vegetable Chains Supplying Hanoi. *Acta Horticulturae* 699: 297–307

Mutersbaugh, Tad. 2005a. Fighting Standards with Standards: Harmonization, Rents, and Social Accountability in Certified Agrofood Networks. *Environment and Planning A* 37 (11): 2033–2051.

Mutersbaugh, Tad. 2005b. Just-in-Space: Certified Rural Products, Labor of Quality, and Regulatory Spaces. *Journal of Rural Studies* 21 (4): 389–402.

Nguyen, Phuong Thao, John E. Bowman, John Campbell, and Nguyen Minh Chau. 2006. Good Agricultural Practices and EUREPGAP Certification for Vietnam's Small Farmer–Based Dragon Fruit Industry. Presented at the USAID Regional Consultation on Linking Farmers to Markets, January 29–February 2, Cairo.

Ong, Kung Wai. 2007. Organic Farming in Asia. In Helga Willer and Minou Yussefi, eds., *The World of Organic Agriculture: Statistics and Emerging Trends 2007*, 107–113. 9th ed. Bonn, Germany: International Federation of Organic Agricultural Movements (IFOAM), and Frick, Switzerland: Research Institute of Organic Agriculture (FiBL).

Organic Europe, n.d. http://www.organic-europe.net/europe_eu/standards.asp.

Phan, Thi Giac Tam. 2006. Vietnam. In Bill Vorley, Andrew Feane, and Derek Ray, eds., *Regoverning Markets: A Place for Small-Scale Producers in Modern Agrifood Chains?*, 125–132. Aldershot, UK: Gower.

Pollan, Michael. 2006. *The Omnivore's Dilemma: A Natural History of Four Meals*. New York: Penguin Press.

Pugliese, Patrizia. 2001. Organic Farming and Sustainable Rural Development: A Multifaceted and Promising Convergence. *Sociologia Ruralis* 41 (1): 112–130.

Ratanawaraha, Chanuan, Wyn Ellis, Vitoon Pinyakul, and Burghard Rauschelbach. 2007. *Organic Agribusiness: A Status Quo Report for Thailand 2007*. Bangkok: Thai-German Programme for Enterprise Competitiveness (Thailand), Sustainable Agriculture Foundation (Thailand), and GreenNet Foundation.

Raynolds, Laura. 2004. The Globalization of Organic Agro-Food Networks. *World Development* 32 (5): 725–743.

Raynolds, Laura, and John Wilkinson. 2007. Fair Trade in the Agriculture and Food Sector: Analytical Dimensions. In Laura Raynolds, Douglas Murray, and John Wilkinson, eds., *Fair Trade: The Challenges of Transforming Globalization*, 33–46. London: Routledge.

Roitner-Schobesberger, Birgit, Ika Darnhofer, Suthichai Somsook, and Christian R. Vogl. 2008. Consumer Perceptions of Organic Foods in Bangkok, Thailand. *Food Policy* 33 (2): 112–121.

Rundgren, Gard. 2007. Certification Bodies. In Helga Willer and Minou Yussefi, eds., *The World of Organic Agriculture: Statistics and Emerging Trends 2007*, 69–74. 9th ed. Bonn, Germany: International Federation of Organic Agricultural Movements (IFOAM), and Frick, Switzerland: Research Institute of Organic Agriculture (FiBL).

Sahota, Amarjit. 2007. Overview of the Global Market for Organic Food and Drink. In Helga Willer and Minou Yussefi, eds., *The World of Organic Agriculture: Statistics and Emerging Trends 2007*, 52–55. 9th ed. Bonn, Germany: International Federation of Organic Agricultural Movements (IFOAM), and Frick, Switzerland: Research Institute of Organic Agriculture (FiBL).

Scott, Steffanie. 2008. Agrarian Transitions in Vietnam: Linking Land, Livelihoods, and Poverty. In Max Spoor, ed., *The Political Economy of Rural Livelihoods in Transition Economies: Land, Peasants and Rural Poverty in Transition*. London: Routledge.

Smith, Adrian. 2006. Green Niches in Sustainable Development: The Case of Organic Food in the United Kingdom. *Environment and Planning C: Government and Policy* 24 (3): 439–458.

Smith, Alisa, and J. B. Mackinnon. 2007. *The 100-Mile Diet: A Year of Local Eating*. Toronto: Random House Canada.

The Nation. 2006. Wangkanai Takes the Organic Route. http://www.nationmultimedia.com/2006/09/04/business/business_30012692.php.

The Nation. 2007. Swift's "Green" Greens Come Home. July 9. http://www.nationmultimedia.com/2007/07/09/business/business_30039897.php.

Thiers, Paul. 2005. Using Global Organic Markets to Pay for Ecologically Based Agricultural Development in China. *Agriculture and Human Values* 22 (1): 3–15.

Vandenberghe, Dirk, and Somprot Sarakosas. 1997. *Capital Rice Co. Ltd. in Europe*. ASEAN Business Case Studies No. 5. Antwerp, Belgium: Centre for International Management and Development.

Vandergeest, Peter. 2007. Certification and Communities: Alternatives for Regulating the Environmental and Social Impacts of Shrimp Farming. *World Development* 35 (7): 1152–1171.

Wiboonpongse, Aree, and Songsak Sriboonchitta. 2007. Thailand. In Bill Vorley, Andrew Feane, and Derek Ray, eds., *Regoverning Markets: A Place for Small-Scale Producers in Modern Agrifood Chains?*, 51–66. Aldershot, UK: Gower.

4

In Whose Interests? Transparency and Accountability in the Global Governance of Food: Agribusiness, the Codex Alimentarius, and the World Trade Organization

Elizabeth Smythe

In a more globalized food system where corporate capital is more concentrated (Clapp and Fuchs, chapter 1, this volume), consumers in many countries are increasingly distant and detached from the sites of crop production and processing. As food production has become globalized, it too has fallen under this complex array of governing institutions and private corporate networks, raising questions about the nature and legitimacy of the rules that are established and whose interests they serve. The voices of consumers and small-scale local producers risk becoming further marginalized in this system. One reaction to globalization and global governance has been a growing demand for increased transparency in the processes of governance. Transparency includes not just increased availability of information and an open process, but also an assurance of accountability of powerful actors, such as governments or corporations, to their own citizens, shareholders, or other affected stakeholders in society.

This chapter focuses on the case of food labeling as a key area of global food governance where issues of transparency and legitimacy arise. The Codex Alimentarius Commission's Committee on Labelling (CCFL) has been working to establish rules on the labeling of foods containing genetically modified organisms (GMOs) since the early 1990s. Thus far, no standard has been adopted, despite much discussion. In order to understand why it has been so difficult to agree on a food labeling standard for genetically modified (GM) food, it is important to examine who influences the development of these rules and to ask some key questions. What interests do they reflect? How transparent is the process? What right do consumers have to information about both the process of how food production and distribution is regulated and the product on the market to be consumed as a result of those rules?

The case of the Codex Committee on Food Labelling highlights the question of corporate influence, along with that of powerful states and other actors in the global governance of food. It addresses the multifaceted nature of corporate power, including instrumental, discursive, and structural facets, demonstrating the array of channels of influence at both the international and state levels. The fourteen-year struggle at the Codex to find mutually acceptable guidelines on labeling foods that have been produced using genetically modified, or engineered, organisms illustrates the role of key actors, including powerful states and corporate agribusiness.

The analysis suggests that the power of these actors, however, has not gone uncontested. Other states as well as consumer and environmental nongovernmental organizations (NGOs), and important norms, in this case, transparency, have challenged that power. As powerful as the transparency norm is, this case also shows that it has not been sufficient at either the national level, in the United States or Canada, or at the international level, in terms of the Codex, to result in standards that legitimate mandatory labeling of GM foods. Powerful actors have thus far successfully opposed a strict labeling standard because it could legitimate rules that might limit product exports.

Foods Standards, International Trade, and the Controversy of Labeling

The controversies around the uses of various technologies in food production, such as the use of growth hormones in cattle or GMOs in various crops, have raised consumer concerns and created pressures for national regulations. Given the globalized food production system, much of it dominated by corporate control of the production chain, and the increasing level of food imports, these developments have implications for states and firms in terms of food trade, access to foreign markets, and corporate profits. The existence of differing national regulatory regimes for food standards and safety and their implications for trade has made the role of international standard-setting bodies in harmonizing regulations that much more important. Since 1995, international standards have also played a role in the resolution of international trade disputes. Thus, what the international standards are, and who influences them, has emerged as a key issue of food governance.

By far one of the most controversial issues has been the issue of food labeling. International food standards, including those regarding labeling, are addressed within the Codex Alimentarius Commission. The extent to which national labeling requirements could be seen as an unjustifiable barrier to trade is thus affected by what goes on at the commission. In that context food-exporting states, corporate actors, and all other actors involved in food production have a material stake in those standards. At the same time, food labels form a primary mechanism by which consumers are informed about the nature of the food they consume, so they too have a stake in the setting of international standards. How these various actors influence the process of international standard setting requires an understanding of the how the Codex Alimentarius operates, how its role has changed, and the way various state and nonstate interests are represented and pursued. These interlocking issues are addressed in the following section.

The Codex Alimentarius and International Trade

The Codex Alimentarius Commission is a joint body of the Food and Agriculture Organization (FAO) and the World Health Organization (WHO) that was founded in 1962. Its mandate was to develop and harmonize food standards, both "protecting the health of consumers and ensuring fair practices in the food trade" (WHO and FAO 2005, 14). A major shift, however, in the significance of its work in relation to trade issues since 1994 has politicized the work of the commission and turned it into a site of struggle around states' right to regulate food and whether such regulations constitute unjustifiable barriers to trade.

Historically, Codex standards were, and continue to be, voluntary guidelines for member states. The guidelines are ultimately, if acceptable to states, incorporated into national food regulations. Since 1995 and the creation of the World Trade Organization (WTO), the standards have been seen by most legal experts as "semi-binding" on states (Victor 1997). This is a result of the conclusion of the Uruguay Round of the General Agreement on Tariffs and Trade, which created the WTO with its more effective dispute-settlement mechanism, and its two agreements on Sanitary and Phytosanitary (SPS) measures and Technical Barriers to Trade (TBT). SPS measures deal with aspects of food safety regulation, while TBT barriers include other regulatory measures adopted to deal with consumer safety, health, or environmental protection that may impact trade. Such measures may include labeling. Indeed, issues related to various aspects of food labeling have been the main focus of the TBT work on trade barriers since 1995 (OECD 2003). Given continued disagreement on labeling, this focus is likely to continue. In using such measures, WTO members are obliged to employ regulations that are the least trade restrictive and, in the case of food safety measures, based on scientific grounds and, where available, international standards.

Codex standards are directly referenced in the WTO's SPS agreement and so could serve as a benchmark, the basis of justification to the WTO, for national measures to protect food safety. Thus national rules that deviate from (i.e., exceed) Codex standards could be seen as illegitimate protectionist measures and become the subject of trade disputes and targets for WTO-authorized and potentially costly trade retaliation, especially for smaller economies or more trade-dependent sectors. Further, the international standards themselves are difficult to challenge at the WTO if they are perceived to be too weak (Buckingham 2000). Standards developed in the Codex can, in essence, reduce or expand the policy space for national food regulation in the name of public safety or other values, such as environmental sustainability. As a result of its changing role, Codex rule-making processes have become more politicized, reflected in its growing membership of state actors (177), the increased involvement of national trade officials pursuing their interests, and the increased attention and involvement of other organizations (e.g., the WTO secretariat). In addition, nonstate actors, both corporations and NGOs, have sought to play a greater role in the standard-setting process both through direct involvement in the work of the Codex Commission and its committees and through efforts at the national level to influence the negotiating positions of state actors (Veggeland and Borgen 2005).

The Codex Commission is in session for two years, culminating in a biannual meeting held in Rome (FAO) or Geneva (WHO). In the interim, work is carried out by an executive committee assisted by a small secretariat drawn from FAO and WHO staff. Much of the Codex work is carried out in committees dealing with functional issues (such as general principles, labeling, limits on pesticide residues) and commodity areas (such as milk and milk products, or meat) as well as a number of issues based on geographic regions. Each member state has a national Codex contact point (often in the national food safety regulating agency), and the national chairs of various Codex committees host the work of the committee—that is, fund the secretariat of the committee and undertake the costs of the committee's annual meeting. Canada has chaired and hosted the food labeling committee's work for many years.

Decisions of the Codex committees and of the full commission are normally made by consensus, but votes can be held in rare circumstances. The development of new food standards follows an eight-step set of procedures that involve the submission of a proposal to develop a standard, a discussion paper, and a decision by the relevant committee that a standard should be developed. A decision of the commission or its executive committee sets out the criteria and a set of work priorities to develop the standard. A draft standard is then developed that may also involve the assistance of a specialized task or work group. Once developed, the draft standard is circulated to all member governments for comment. Depending on those comments the draft may be revised and ultimately adopted. The process can take years. Given the increasing demand for, and complexity of, food production and standards and the small size of the secretariat, coupled with the increased politicization of the organization since 1995 (Veggeland and Borgen 2005), the process of developing new standards is taking longer than ever.

The Representation of Interests in the Codex and the Extent of Corporate Influence

Policies on the representation of various interests in food production at the Codex have traditionally allowed for more input from nonstate actors, especially food producers and processors, and more transparency, than is the case for other international organizations like the WTO. This relative openness has, given the changing trade significance of Codex standards, provided a direct channel for corporate influence over the development of international standards.

According to the Codex *Principles Concerning the Participation of International Non-governmental Organizations* (INGOs) elaborated in 1999, any organization that has an international structure and scope of activity (in three or more countries) can be accorded status and participate as an observer, receive all documents, and speak at the discretion of the chair (normally after interested country delegations have spoken). In addition, many national delegations include representatives of various NGOs or industry and corporate associations in their delegations for full commission and committee meetings. As many critics point out, large corporate agribusiness has a vested interest in uniform standards and global market access, reflected in its participation in Codex work.

A 1993 report indicates that of the total of 2,578 participants on the various Codex committees, over 660 represented industry (National Food Alliance 1993). The commission at that point included 105 national delegations, while 140 corporations were also represented. This trend has continued. The meetings of the commission in 2000 and 2002 included about 150 of these INGOs (Codex term). Of these about 70 percent represented industry (WHO and FAO 2002; Sklair 2002). By 2005, the number of INGOs had reached 156, and for the June 2007 meeting 157 are listed. The number of observers has, in fact, increased more rapidly than state membership (Huller and Maier 2006).

The Codex Committee on Food Labelling has followed a similar pattern. In its May 2000 meeting there were 50 INGOs represented of which 35 were from industry and about 15 from consumer or other groups. In meetings of the committee in May 2006 and 2007 in Ottawa the trend has continued. About 20 of the 25 observers in 2006 appeared to represent industry and in May 2007, of the 27 present 21 appeared to represent producer or corporate organizations. Moreover, the composition of national delegations often includes industry representatives and a few other organizations. Of the 200 official national delegates in 2000, 48 came from industry (Sklair 2002, 164) while only 4 represented consumer organizations (Consumers International 2006). In the case of the most recent committee meetings on labeling, for example, Canada's delegation included the umbrella organization BIOTECanada ("Canada's voice for biotechnology"), represented by a Monsanto executive, along with representatives of corporations such as Kraft, Nestlé, and Mead Johnson.

In general, corporations and industry associations are better able to fund the participation of a larger number of representatives at various Codex meetings. Even with their more limited resources, other consumer and environmental NGOs use their access to committee and commission meetings to disseminate reports of meetings through the extensive use of the Internet. However, the lengthy, complicated process of the development of standards across numerous committees requires a long-term commitment of staff with expertise. This is a challenge for all but a few NGOs to sustain.

In addition to the imbalanced representation of interests in terms of observers and nongovernmental representatives included in official delegations, national participation is also highly imbalanced, despite the increased state membership in the Codex. In this aspect it is more similar to the WTO. Developing countries, especially the least developed countries, often do not participate in meetings or send small delegations of a single individual, in contrast to countries such as Canada, which often sends over twenty people to a commission meeting, or twelve to fifteen to a committee (author's calculations). Recognizing the challenges for developing countries, the Codex created a modest trust fund in 2003 to enhance participation and, in 2004, funded ninety officials from developing countries (WHO and FAO 2005, 35). Developed countries continue to dominate as well in their role as chairs because of the costs of hosting Codex committee work and the limited budget of the secretariat. Even with the trust fund, participation in Codex work remains very imbalanced in favor of Northern countries. Since 1995 the politicization of the Codex and its relevance to trade has also been reflected in the increased use of trade representatives in national delegations and closer involvement of the WTO Secretariat.

While direct participation in the Codex process as observers or through official delegations is one way corporate and other groups can seek to influence Codex standards, the other is, of course, through influencing national food regulations and the national positions the most dominant state actors adopt on Codex matters and trade policy. The domestic context, along with negotiations at the Codex, therefore, is especially relevant to understanding corporate instrumental power and whose interests are reflected in international rules, and will be dealt with further below. I turn now to the work of the Codex and the case of the Codex Committee on Food Labelling.

The Codex, Science, and Power

With changes to the status of Codex standards in 1995 has come a changing attitude on the part of many state and nonstate actors and greater willingness to use the Codex to advance trade interests. The United States, for example, began to more actively block the adoption of more stringent food safety standards, particularly in areas where there was potential scientific dispute (Thomas 2006).

Similarly, the European Union has sought to use food diplomacy to advance its interests and to block Codex standards, for example, in the case of bovine growth hormones when the emerging standard did not support EU regulatory practice. When that effort failed, the EU became the subject of a WTO challenge over its ban on US and Canadian beef. On the other hand, when US attempts to gain acceptance of the use of synthetic hormones to increase milk production via a Codex standard also failed, the basis of another trade challenge against the EU disappeared.

What had happened in the meantime is partly a reflection of the failure of the United States to alter European acceptance of GM foods and a broader difference in the basis of regulatory systems. Since 1998, Europe had not approved any new GM products, under pressure from member countries demanding it not do so until labeling and traceability[1] rules were put in place. On July 2, 2003, the European Parliament approved two laws that required the labeling of GM products. Given the negative impact of the GM moratorium on food exports, the United States (June 2003) and then Canada (August 2003) launched a WTO trade dispute. The EC–Beef Hormones case had also been launched by Canada and the United States. In that case the complainants won, but the victory did not result in the opening of the EU market to hormone-fed beef, since the EU was willing to pay costs of compensation rather than bear the wrath of EU consumers. Like the beef hormone case, the May 2006 WTO victory over the EU on the GM moratorium again may prove to be pyrrhic since it is unlikely, given the strength of consumer concern in Europe, that regulatory policies will change.

In each case, a central issue has been the role of the scientific justification in terms of food safety and the role of the precautionary principle in contrast to risk assessment and risk management. The role of scientific knowledge and uncertainty is relevant to understanding the discourses around the labeling of GM foods and their link to various forms of discursive and structural power.

The scope of risk assessment within the Codex has been restricted to human health in various foods. Given its small secretariat and limited resources, the Codex relies heavily on various "independent experts" for scientific advice on questions of health risks. This combination means that certain knowledge and rationales for setting and regulating food standards are acceptable, while others are not. Although the Codex does allow for "other legitimate factors" to enter the process at the risk-management stage, these have been the subject of great dispute, especially within the Codex committee on General Principles, and, in the case of trade disputes, have often been defined in terms of economic or trade issues (Huller and Maier 2006, 288). Where scientific uncertainty exists, or important social factors intervene, such as consumers or environmental concerns, the resulting differing national regulations can form the basis of trade disputes.

The Codex Committee on General Principles establishes the formal rules and procedures for the Codex over the various committees and their operations. The committee agenda for almost a decade has included discussion of the principles for risk analysis and food safety. It is here, along with other Codex committees such as Labelling, and within WTO bodies that the European precautionary-based regulation and US science, or risk-based regulation, have come into conflict. The European Union has pushed hard to ensure that precaution—the notion that in cases of insufficient, inconclusive, or uncertain scientific evidence governments may take preventive or precautionary regulatory measures—is reflected in the principles. The United States and its allied biotechnology and food industries have battled hard to keep references to precaution out of the principles. To date they have been successful, but the debate also has embedded within it various material interests of actors. Thus privileging independent scientific assessments of safety has implications for the discursive and structural power of various actors. The globalized nature and increasing scale of food production, along with the rapid development and changes in the technology of food production, especially biotechnology, means that those with the scientific resources and knowledge have advantages within this discourse over those who lack such resources—a major issue that many social movements confronting aspects of agribusiness and sustainability have faced.

Codex procedures of risk assessment have typically relied on WHO and FAO experts, such as the Joint Expert Committee on Food Additives, the Joint Meetings on Pesticide Residues, and the Joint Expert Meetings on Microbiological Risks. Experts review the evidence contained in existing studies, most of which are furnished by the "food manufacturer's scientists, rather than conducting first-hand empirical studies of their own" (Huller and Maier 2006, 284). In terms of the work of the Codex, the issue of scientific advice and expertise has been one of the areas, reflected in an extensive evaluation of the work of the Code in 2002, that have been seen by critics to be the least transparent of its processes of operation (Suppan 2006).

What is independent, disinterested scientific knowledge is not always clear. Corporate actors have sought to enhance their authority and legitimacy on controversial issues around food, or other product safety, through the creation of what Buse and Lee (2005, 13) have called "institutionalized non-profit industry established and funded scientific networks" such as the International Life Sciences Institute (ILSI), which claims to be "a global network of scientists devoted to enhancing public health decision-making" (ILSI 2007a). But the organization was founded in 1978 by various food and beverage firms, including Coca-Cola, and had links to the tobacco industry (Sell 2007). It has extensive links to the FAO and is very active in the work of the Codex, including the Committee on Labelling. The ILSI Food Biotechnology Committee works to "support the development and harmonization of science-based regulations in the world for biotechnology-derived food products and disseminate scientific information regarding the safety assessments of these products to governments, industry, and other interested groups" (ILSI 2007b). ILSI has strongly opposed the mandatory labeling of GM foods.

The privileging of science-based risk assessment of food hazards over other factors in both the Codex and the WTO, and over the growing consumer concerns in some countries, in the view of some observers, will to lead to more politicization of the Codex and more trade disputes (Thomas 2006). Deadlock has indeed characterized the long, slow saga of food labeling in the Codex. The outcome to date at the Codex, however, has not wholly reflected the preferences of powerful actors like the United States and allied biotechnology and food processing corporations. Rather, the case suggests that other actors and other forms of power are also relevant. As the following discussion illustrates, the United States sought to achieve standards in the Codex that did not mandate the labeling of GM foods, and when that effort failed, moved to block further work on standards and then pushed for their abandonment, so far without success. On the other hand, consumer concerns, and powerful norms of transparency, including the right to know, have not yet proven sufficient to achieve a Codex standard of mandatory labeling.

Codex Committee on Labeling

National positions on labeling GM food are driven, in part, by economic interests of GM crop producers. As such, the structural power of corporate actors involved in the production of GM crops and food within these countries varies. By 2004, it is estimated that 81 million hectares of GM crops were planted globally, in contrast to 1.7 million in 1996 (Pfister 2005). Crops include soybeans, maize, cotton, and canola developed for herbicide tolerance and insect or pest resistance. GM crop production is heavily concentrated in a few countries, including the United States, Argentina (Newell, chapter 9, this volume), Canada, Brazil, and China. Given its early role in the development of biotechnology and its strong support for the industry, the United States has a strong material interest in GM food. Both the Grocery Manufacturers of America (GMA) and Greenpeace agree that 70 percent of US food in supermarkets contains GM ingredients. European countries, in contrast, have been more divided on biotechnology, slower to develop GM products, and faced much stronger consumer and environmental opposition. Corporate food retailers responded to consumer demand in limiting or eliminating GM foods from their supermarket shelves (Bernauer 2003). These differing interests and levels of influence of the biotechnology industry are reflected in the development of domestic food regulatory systems, and national positions on labeling in the Codex, as the following overview of the CCFL negotiations reflects.

In 1991 the Codex Commission agreed on the need to address the issues of biotechnology and GM foods, and the CCFL agreed that work on labeling aspects of biotechnology should begin. In April 1993 the United States, as the leading country in the industry, was asked to prepare a discussion paper that was initially discussed in the October 1994 session. Debate centered around whether labeling should be required only when there were health and safety concerns and whether labeling should be required if the foods in question did not differ substantively from traditional equivalents. However, for many country delegates, including those of Canada and the EU, the issues were premature since they were still reviewing and developing domestic regulations. Consumer groups—in this case, Consumers International (CI)—favored a system of comprehensive labeling based on the consumers' "right to know." Others also argued in favor of labeling that indicated how food was produced in order to permit consumers to make choices based on values other than just those of health and safety. In the absence of a clear consensus the issue was ultimately referred back to the commission's executive committee.

By April 1997 the secretariat had produced a set of Draft Guidelines based on previous work, but after delegate complaints about the short timeframe in which to consider the guidelines, the committee decided to take more time to solicit member comments. The guidelines would have limited labeling for those GM foods that were not considered equivalent to traditional foods. There were also specific proposals on labeling in relation to allergens. This more restricted approach to labeling was supported by the country delegates of the major producers of GM foods, which included the United States, Brazil, and Mexico, along with the major corporate players in the biotechnology industries. Norway advocated a broader approach that reflected the right of consumers to know and choose, supported by consumer organizations. These divisions would be replicated in subsequent meetings of the CCFL as efforts to find a consensus became ever more elusive.

In 1999, an alternative to the first set of draft guidelines had emerged that would allow for all foods containing GMOs to be labeled. Consumers International supported this more inclusive approach. In opposition, the United States and Argentina made the argument that labeling was unnecessary, given the equivalence of GM foods to conventional foods. The United States raised the concern that labeling based on the method of production would imply that GM foods were unsafe and would deter consumers. The United States was supported by a number of industry associations. In the absence of consensus, again the committee opted to create a working group, coordinated by Canada, to rewrite the draft and develop the two options. By 2001, the working group's revised draft now included three labeling options. Despite the optimism of the Canadian chair that progress, though slow, would lead to a consensus, it proved to be illusory (MacKenzie 2001). By 2003, the committee acknowledged it had made little progress on the labeling guidelines and another working group was established to develop options. Their report was reviewed in the 2004 meeting.

The continued US opposition to labeling based only on the "method of production" rested on the argument that such a policy would constitute an unfair trade practice since consumers would perceive the label as a safety warning. Thus, such labels would constitute a market barrier. The United States argued that only cases where significant changes in the product composition had occurred were legitimate candidates for mandatory labeling. Canada concurred and also reiterated the US claim that developing countries would be unduly burdened by broader labeling guidelines. Not surprisingly the European Union, which had just developed its own labeling and traceability regulations in 2003, and had been subjected to a US and Canada trade challenge on its earlier moratorium on GM approvals, opposed the US position.

Since the inception of this work at the Codex, the US position has, in fact, lost ground as more countries have opted to develop some system of labeling that goes beyond the US position. By 2005, countries supporting a more comprehensive labeling of GM food included the EU countries, China, Japan, Korea, Thailand, India, Nigeria, Kenya, Cameroon, Malaysia, Australia, and New Zealand. Those nonstate actors on the comprehensive labeling side included Consumers International, the International Federation of Organic Agriculture, Greenpeace, and the Erosion, Technology and Concentration (ETC) Group. Those favoring very limited labeling included the major biotechnology organizations such as CropLife, the Biotechnology Industry Organization (a US industry advocacy group), BIOTECanada, the International Association of Plant Breeders for the Protection of Plant Varieties, and the International Council of Grocery Manufacturers (US PIRG 2005).

By May 2006, the preferred position of the United States and biotech industries was for the Codex to abandon the search for guidelines on labeling altogether. This was preferred over bringing forward recommendations for mandatory labeling that would be more comprehensive and would cover any products containing GM components, whether or not they differed from conventional counterparts. Moreover, many smaller countries had lined up behind the bigger opponents of a liberal view of GM regulation (such as the EU and Japan) largely because they feared the trade implications for their own exports in these markets if they accepted GM products without labeling or traceability. In the context of the strong resistance already to GM food, any development of mandatory labels at the Codex would merely limit the United States and the biotech industries' ability to push for more market access via the WTO.

At meetings of the CCFL in 2006 and 2007 the major issue once again was GM food labeling. The United States and its biotechnology allies lobbied vigorously to suspend CCFL work on GM labeling in 2006, at one point characterizing it as a waste of time, given that there was no consensus or the likelihood of one emerging and that the committee's resources would be better used in other areas (Codex 2006). The Canadian chair also seemed willing to entertain the possibility of a suspension, at least until there was new evidence as a basis to reopen the discussion.

However, proponents of labeling, supported by consumer organizations, intervened, and led by Norway, they proposed establishing another working group to continue the work on guidelines on mandatory labeling. Norway argued that labeling was vital to ensure "that consumers were able to make informed choices" (Consumers International 2006). In this they were supported by nineteen countries and the EC. Representatives of Consumers International were also included as members of national delegations from India, Morocco, and Burundi. Ultimately the proponents of mandatory comprehensive labeling were able to keep the process alive.

The committee met in Ottawa in May 2007, heard the report of the working group, and reviewed seven different approaches to the labeling of biotech foods ranging from the EU' s mandatory approach to Canada's purely voluntary one. Those opposed to mandatory labeling included the biotechnology industries, the United States, Canada, Australia, and New Zealand, which argued that labeling based "on the method of production was not justified on the grounds of food safety or fair trade practices and that the consumers' right to know was not one of the objectives of the Code" (ICTSD 2007). In an ironic twist, one observer from the 49th Parallel Biotechnology Forum, who disagreed with this argument, stated that "a large proportion of biotech foods being sold have not been subject to *any* safety assessments and therefore labeling helped consumers make their own decisions about health and safety" (ICTSD 2007). Consumer organizations continue to make this argument that the issue of labeling, given the weak regulation and scientific uncertainty around GM food, cannot be separated from questions of safety (Consumers International 2006).

In the subsequent discussion, a group of countries and observers (called the Friends of Mandatory Labelling) argued that work on labeling should continue, that the Codex did have a mandate to deal with the issue, and that consumers' right to know and make informed choices was an essential element of labeling. Many developing countries felt the commission did have an obligation to provide guidance on this issue.

Once again, the United States, Argentina, Canada, Chile, Australia, and New Zealand, along with the biotechnology industry representatives, argued in favor of discontinuing work on the issue given members' major differences over labeling approaches. Ultimately the committee decided to continue work on labeling guidelines. Another working group was formed to "examine the rationale for adopting or not adopting a particular labeling approach, as well as strategies to communicate to the public on GM foods" (ICTSD 2007). In addition, the working group would examine whether existing Codex texts supply sufficient guidance on labeling, as the United States claims they do.

As the struggle over labeling at the Codex suggests, the extent of corporate power in food governance and the creation of international standards and trade rules cannot be fully understood without examining the channels of influence at the national level, given that it is still state actors that make the decisions regarding those standards and trade rules. Thus the positions member states take on labeling at the Codex reflect policy struggles within these countries and the power of various actors—be it structural, instrumental, or discursive. In the case of the United States, one of the most powerful state actors at the Codex, and in Canada, which has chaired the CCFL for many years, consistent opposition to mandatory labeling at the Codex reflects the instrumental, structural, and discursive power of corporate interests, especially in the biotechnology and food manufacturing and processing industries. However, there have also been limits to their discursive power. These countries are also ones where there are strong emphases in the culture and governing institutions on transparency and on the welfare of the public, especially the "consumers' right to know" (DeBoer 2003). Consumers in Canada and the United States, according to polls, have also consistently supported labeling of GM foods.

Industry and Consumers in the Domestic Politics of Labeling: The Need to Know vs. the Right to Know

The politics of food regulation and labeling reflect the national context and the power of various interests and actors. Some observers see a "regulatory polarization" between the European Union and North America and different national regulatory cultures, satirically characterized as follows:

In the US products are safe until proven risky
In France products are risky until proven safe
In the UK products are risky even when proven safe
In Canada products are neither safe nor risky.
(Bernauer 2003, 9)

The reality, of course, is a more complex interplay of institutions, values, actors, and interests. In the case of the EU food scares, factors like early resistance to GM products, less investment in GM food crops, business conflict (Falkner, chapter 8, this volume), a strong NGO mobilization, and eroding consumer trust in regulators all played a role in creating a tighter regulatory environment. In contrast, both the United States and Canada were among the earliest adopters of biotechnology and vigorously supportive of the industry and GM crops—facts that partially explain the strong structural and instrumental power on which the biotechnology and food manufacturing industries have been able to draw (Kuyek 2002).

In both Canada and the United States, the embracing of the biotechnology sector came early and involved close links between the biotechnology industry, government departments, and regulatory agencies. In the case of the United States, the US Department of Agriculture (USDA) saw biotechnology in it early stages as a "frontier" technology and one in which the United States clearly had a competitive advantage (Prakash and Kollman 2003, 620).

The instrumental power of the US biotechnology industry—deriving from its extensive lobbying, resources, and access to US administrations, policymakers, and regulatory authorities—has been well documented. This includes the revolving-door phenomenon of the movement of regulatory and other officials from the private sector into the public, or vice versa. The movement and personal connections between biotechnology corporations such as Monsanto, the USDA, the Food and Drug Administration (FDA), the Environmental Protection Agency, and the Trade Representative, during both the Clinton and George W. Bush administrations, are striking (Smith 2003). In addition, biotechnology corporations have been generous funders of US election campaigns, with their contributions totaling US$3.5 million between 1995 and 2002 alone. The major advocacy organization—the Biotechnology Industry Organization—spent US$143 million on lobbying from 1998 to 2002, which included regular meetings with, and support for, the biotechnology caucus of Congress to "educate" them on the industry (Charman 2001).

However, this instrumental power alone does not account for their broad influence over regulations for labeling GM food. Their structural power, in terms of significance to the US economy, is also complemented by significant discursive power reflected in their links to the scientific and university communities and their role in shaping and framing the debate over GM food labeling, especially as it relates to questions of safety and costs to consumers.

Similarly in Canada, after the embrace of intellectual property norms and the abandonment of generic drug companies, part of the continental integration reflected in the Free Trade and NAFTA agreements, the Canadian government saw biotechnology as a key "cutting-edge" sector. The Canadian Department of Agriculture and Agri-Food, and the Department of Industry, sought to stimulate the development of the industry and attract more foreign investment. The movement of the Canadian Food Inspection Agency (CFIA) from the Department of Health to Agriculture Canada in 1997 also marginalized its role in food regulation and gave Agriculture the key (and somewhat conflicting) roles of both promoting and regulating the industry. Although the revolving-door phenomenon and campaign finance did not play much of a role in Canada, strong government support for the biotechnology industry ensured that regulation would be fairly minimal, as would the development of an independent public-sector research capacity. In both countries close links between biotech companies, university research institutes, and government went on in the early years. This resulted in what has generally been characterized as weak or limited regulation of biotechnology, much of which is largely outside the public purview. However, this was not to remain the case in either country.

As Barrett and Abergel (2000, 4) point out, an important aspect of what they call the deregulation of GM products in North America is linked to the development of a regulatory discourse around familiarity and substantial equivalence developed in the 1980s and early 1990s by industry advocates, scientific agencies (the US National Research Council), and the OECD, and then further elaborated by Canadian regulators. The concept of familiarity refers to the experience with a chemical or GM crop on a small scale, which could be extrapolated to define and foresee risks of commercial release, while substantial equivalence assumes that a new GM product can by compared systematically to existing, natural counterparts that have a history of safe usage. If they are found to be familiar and substantially equivalent, such products require no special regulatory treatment. This facilitated rapid commercialization and international exchange and helped to marginalize skeptical and critical voices and "exclude the public from risk debates" (Abergel and Barrett 2000, 2). As we see below, it also formed part of the discourse around food safety and the need to label GM foods. The marginalization of skeptics was, however, incomplete and critics of the limited regulation of GM technology, armed with the norm of transparency, did demand greater input into the regulatory process and information about GM food in both Canada and the United States.

In the US case, according to Prakash and Kollman, the shift occurred around 1999–2000 with the marked increase in court challenges relating to the marketing and release of GM products into the environment and increased congressional attention to GM regulation in the form of over fifty bills, a number of them dealing with labeling. In 2000, the StarLink food scare, where unapproved corn products ended up in taco shells and forced a recall and later a farmers' class action lawsuit against Aventis, led to NGO protests and much more media attention to GM foods. Consumer confidence and concern had increased, as had the demand for labeling. Polls showed 75 percent or more of the public wanted labeling (Prakash and Kollman 2003, 631). Pressures developed to address these concerns and the perception that regulators were only listening to corporate interests.

In 1999, the FDA responded to concerns by holding a series of public meetings in Washington, Chicago, and Oakland, California. It claimed that its goal was to share the FDA approach regarding safety with the public, to gather views on whether policies needed to be modified, and "to assess the most appropriate means of providing information to the public about bioengineered products in the food supply" (Da Ros and Lynch 2001, 14)—in essence, to address the issue of labeling. The meetings were open to all, but began with expert panels and lengthy presentations with public input limited to brief comments at the end of the day. Public interest was high with over 1,000 in attendance at each meeting. The meetings reflected both divisions among experts on many points and a high level of public concern about information. Yet, as Da Ros and Lynch (2001) point out, many views of the public that did not accord with the FDA were not reflected in the FDA recommendations in 2001 on voluntary labeling.

Concern had also been evident at the national level with congressional bills, such as those introduced by Dennis Kucinich called the Genetically

Engineered Right to Know Act,[2] and in a number of states that have sought to regulate GM crops or food labeling. In Oregon the 2002 election included a ballot initiative calling for the mandatory labeling of GM foods. In that case, however, the industry was able to turn public support prior to the election around partly by outspending the supporters of mandatory labeling by US$5.5 million to US$160,000 (US PIRG 2005).

While Prakash and Kollman (2003) claim that these developments created the political space for change in the US policy position on GM food regulation, such changes cannot be seen on the international side in its position at the Codex. What did become institutionalized were formal consultations with various interests prior to Codex meetings and an annual public meeting in downtown Washington for three hours late in the afternoon, about one month before meetings of the CCFL. To date, powerful corporate actors in the United States and the retail and grocery associations (with some exceptions in the fast-food and baby-food industries) have continued to favor only voluntary labeling. Along with the US government and the biotechnology industry, they remain strongly opposed to mandatory labeling at the CCFL. Opposition to labeling at the Codex and concerns about European policies on labeling have been communicated loud and clear to US negotiators over the years. Robert Harness of Monsanto, speaking before a congressional subcommittee in 1998, made that clear:

> The second trade issue is product labelling which is a matter of ongoing debate in Europe and one which has significant potential to impact trade. In the United States, food labels are used to inform consumers of nutritional and safety differences in foods they eat. Foods produced through biotechnology are not singled out as a specific category for labelling. The policy debate in Europe is based on the use of the food label to inform consumers of the presence of a "product of" biotechnology, regardless of whether the food is different from a safety or nutritional standpoint.
>
> Monsanto fully supports the science-based approach to food labelling used by the US FDA. While we recognize that local labelling programs can be used to provide information to consumers, as is the case in Europe, such "information" and process labelling must not be allowed to create non–science based segregation requirements in commodity crop markets that will lead to trade disruption and significant cost increases for consumers. (Harness 1998)

The themes of a science-based approach, voluntary labeling, substantial equivalence, and the potential cost to Americans, either in exports or higher food prices of mandatory labeling, are clear, as is the refusal to acknowledge any basis for the consumer's right to know the process by which the food was produced.

Allied with the biotechnology firms have been the major manufacturers of food in the United States. The merger of grocery and food products associations created, as their website indicates, an association "comprised of fifty-two chief executive officers from the Association's member companies. The US$2.1 trillion food, beverage and consumer packaged goods industry employs 14 million workers, and contributes over US$1 trillion in added value to the nation's economy" (GMA 2007). The manufacturers too have had a largely consistent position on the issue of GM food labeling, which they reiterated in a letter to the chief US delegate to the CCFL in 2003:

> Generally the GMA continues to urge the United States Government to categorically oppose any Codex draft recommendation on labelling of foods obtained through techniques of modern biotechnology that is not wholly consistent with the Food and Drug Administration's science based labelling policy. Consistent with the mandate of Codex to protect the health of consumers, GMA endorses a Codex labelling standard for such foods that may contain a potential allergen and/or are materially different from conventional crops in nutritional values. While many Codex members and observers support an International Codex Standard on labelling which would identify the process in which an ingredient is made, there is no scientific basis for such a Standard. However, such labelling is appropriate on a voluntary basis governed by market incentives. (GMA 2003)

In Canada growing demand for more regulation also put pressure on policymakers to be more transparent about the regulatory process, but has not yet proven sufficient to result in mandatory labeling. Concerns about GM food and labeling have been repeatedly reflected in public opinion polls in the last seven years. Two private members' bills on mandatory labeling were introduced in the Canadian House of Commons in 2001 and 2003, the first only narrowly defeated at second reading and resulting in a hearing of the Commons Standing Committee on Agriculture. The response of the Canadian government has come primarily in the form of limited transparency and consultations. In September 1999 it created the Canadian Biotechnology Advisory Committee (CBAC) with a mandate to provide "expert advice to the federal government on the ethical, legal, social, regulatory, economic, scientific, environmental and health aspects of biotechnology." Its twelve members are technical experts drawn from academia, along with some industry, consumer, and environmental representation. It reports to the Biotechnology Ministerial Coordinating Committee, whose members include the ministers of Industry, Agriculture and Agri-Food, Health, Environment, Fisheries and Oceans, Natural Resources, and International Trade. In addition, the government has established a biotechnology gateway website and a transparency project to solicit submissions on applications for approval of new GM products.

Questions, however, have been raised about the relationship between regulatory agencies and the industry and the extent to which public resources have gone into selling the public on GM foods rather than into regulation. Documents obtained under the Access to Information Act by an investigator with a Canadian NGO, the Canadian Health Coalition, indicated a growing unease on the part of the industry and the Canadian government over the campaigns of organizations like Greenpeace and the Council of Canadians. This resulted in a 1999 private meeting of the Minister of Agriculture with the CFIA, the industry lobby group BIOTECanada, and a representative of something called the Food Biotechnology Network, a supposedly neutral provider of "balanced, science-based facts about food biotechnology and its impact on our food system" (Stewart 2002). The group, in fact, had received federal funding to create and produce pro-biotechnology materials for use in schools and by dieticians. The Consumers Association of Canada (CAC) was also enlisted in the government's effort to counter the growing concern over GM food partly because it was vulnerable to co-optation, being in dire financial straits and in need of funding (CAC 2003). The CAC spokesperson later joined Monsanto. Together these groups under the coordination of the public relations firm Hill and Knowlton formed a task force and a national coordination strategy, which included a glossy insert in a national homemakers' magazine, *Canadian Living*, promoting GM foods under the ironic title "A Growing Appetite for Information" (Stewart 2002). The CFIA later sent every household in Canada a brochure on GM food. However, internal polls done for the CBAC and other agencies continued to show public concern and a desire to know which foods contain GMOs.

These efforts did not end the concern about GM crops and food. In fact some observers argue that the positions if anything became more polarized as anti-GMO networks were formed and were able, despite the strong support of business, government, and scientists for the technology, to raise concerns. This was partly because there was no obvious benefit to consumers from GM food, and the tight association of government, industry, and those in the scientific community benefiting from the biotechnology industry created a "legitimacy asymmetry gap" that worked to the benefit of the NGOs in the debate over GM foods (de Clercy et al. 2003). Even in provinces like Saskatchewan, where the industry has a strong presence, an Eco Network had formed and, along with other groups, boycotted the CBAC, seeing it as a cynical attempt at co-optation.

In 2003, the government mandated the Canadian Standards Board to develop standards on labeling. The standards developed were so weak that under pressure from its members, the CAC—which had been part of the consultative process—pulled out, claiming that the standards "do not adequately represent consumers' concerns" (CAC 2003). They noted that over 95 percent of consumers demanded labeling. The association demanded that the labeling be mandatory and that the threshold for the presence of GM products be lowered from 5 percent to 1 percent, the prevailing standard in other major countries, or to the recently announced EU standard of 0.9 percent. In the wake of the voluntary labeling policy, few food producers have labeled their products. The new standards have not eased public concern, as a 2004 Leger Marketing poll released by Greenpeace, *Option consommateurs and L'Union des consommateurs*, indicated on the eve of the CCFL meetings in May 2004 in Montreal. The poll showed that 88 percent of Canadians (over 91 percent in Quebec) wanted mandatory labeling. Over 80 percent of respondents felt that the government had not provided adequate information on GM food so they could make an informed choice (Greenpeace, Option consommateurs, and l'Union des consommateurs 2004).

One government response to concerns about GM food was to focus the discourse on safety. In an effort to show the safety of these foods and their regulation, the government had called on the prestigious Royal Society of Canada (RSC) to put together an expert group to review its regulatory process and make recommendations on regulation. The report, however, raised further questions. First, it indicated that much of the existing regulatory regime accepted the idea of "substantial equivalence" discussed above, which assumed that if the components of the GM product were the same as those already deemed safe, the product would, in its entirety, also be considered safe. In the view of the RSC such conclusions had little scientific justification. Second, the report pointed to a close relationship between the industry and government and the excessive secrecy of the regulatory process. Part of the government response has been the efforts at transparency outlined above.

Canada too has had court cases that have raised the profile of the debate over GM crops. The most celebrated was that of Percy Schmeiser, a seventy-year-old Saskatchewan farmer whose crop was contaminated by a Roundup Ready canola seed. The seed, developed by Monsanto, was sold under license as resistant to the herbicide Roundup—also sold, of course, by Monsanto. When Schmeiser replanted saved seeds from his farm the Monsanto Corporation sued him. The case dragged on for seven years with the lower courts finding him guilty and awarding legal costs to Monsanto. Schmeiser appealed all the way to the Supreme Court and became a folk hero to those concerned about GM crops, contamination, and seed patenting. In a narrow decision, the Supreme Court of Canada upheld the judgment in favor of Monsanto but overturned the award of costs to Monsanto.

Despite consumer concerns, Canada has not moved to mandatory labeling nor has it significantly altered its opposition to such labeling in the CCFL. The official position at the Codex reflects, again, the preferences of the biotech industry as outlined by BIOTECanada and Monsanto. This situation illustrates the extent to which the biotechnology industry has used its access to the Canadian government as a channel of influence in global food governance, along with its structural power based on its role in the economy and the importance of GM crops. Its discursive power to frame the issues of food safety in terms of substantial equivalence and sound science has not gone unchallenged within Canada, partly because of the strength of norms and ideas around transparency of information and the consumer right to know.

Power and the Case of Labeling

How power shapes ideas and norms, frames issues, and determines which ideas and what knowledge are considered valid is also important in understanding the struggle over the governance of food and especially the issue of labeling foods. Major GM-crop-producing countries (such as the United States and Canada) and the biotechnology industry have questioned the need to disclose the GM origins of a food product. They privilege "scientific expertise" on the safety of the components of GM food and argue that they are substantively equivalent to traditional components. Given that fact, they claim, there is no need for consumers to be informed or indeed for any rigorous system of regulation and approval of the resulting food products. Many officials and company representatives use the argument that their techniques are merely an extension of plant breeding, a practice going back thousands of years. The assertion that there is nothing new here is, of course, at odds with the strong demand for tight regulation of intellectual property on the part of the biotechnology industries (Sell, chapter 7, this volume) to protect their novel products! Opponents also suggest that labeling, given the extensive presence and use of GM products as inputs, would lead to massive costs to ensure traceability and crop segregation. All of which would be passed on to consumers. However, as our discussion above has indicated, scientific experts are not in full agreement on these "facts."

The desire for information on the method of production as a basis for consumer choice, grounded in some other value such as environmental, moral, or ethical concerns about GMOs, is also questioned. Any basis other than science-based safety issues (themselves narrowly construed, as the RSC report pointed out) is deemed illegitimate. Moreover, in trade terms, any regulation not based on the safety and science nexus is argued to be "unfair" and thus an unjustifiable barrier to trade.

An additional argument has also been made against consumers' right to information that again privileges expertise. This is the claim of the industry and some experts that ordinary consumers are not capable of comprehending the complexity of GMOs, so that the information conveyed on food product labels would only confuse and alarm them. William Leiss quotes Phillips and Isaac, who argue that such labeling would not be in the public interest because misinformed and confused judgments of some consumers, on seeing a GM label, might lead them to reject the product. As Leiss (2003, 7) points out, the idea that a few misinformed "judgements about GMOs . . . constitutes a justification for industry to refuse to identify the products it makes using gene technology" makes no sense and could not be "generalized to any other areas of life." Yet this is exactly the argument that industry has made, as have the United States and other opponents of labeling at the CCFL. There are other dangers to this view of assuming that the public is irrational, because it will lead to little effort to inform the public, which can produce anger that decisions are being made without meaningful public input (Hallman 2000, 18).

In the broader debate over GM foods both sides have also appealed to other norms, including those of environmental sustainability and poverty reduction. The biotechnology industry claims that its crops minimize pesticide use and thus have a positive environmental impact. Claims of pest-resistant, higher-yielding, and more nutritious crops are also part of the industry's discursive strategy to push GM food production in the global South, particularly in Africa (Williams, chapter 6, this volume). In contrast, opponents point to threats to biodiversity, the risks of monoculture, and the growing corporate control and patenting of life forms. While these broader ideas cannot be discussed here, they form an important context within which the debate over labeling occurs.

Still, the norm of transparency and the right of consumers to information and to know more about the food they consume remain powerful. The demand for more transparency of information has not diminished in terms of food labeling. The resistance of industry and government to labeling is seen as suspicious, perhaps suggesting there is something to hide and therefore, a basis for concern about these products. Even some who support GM foods are not opposed to more regulation and have argued that only by mandatory labeling will the industry and regulators be able to build trust with consumers.

The biotechnology industry has felt the demand for more transparency in labeling, which in some regions of Canada, especially Quebec, continues to be strong. Their response also draws on the structural power of their industry, as a representative of CropLife Canada (the association of the largest biotechnology firms) showed in his submission to the Quebec National Assembly Commission on Agriculture's inquiry on food labeling in 2004. He pointed out first how many jobs (over 31,000) and how much research and development investment (US$349 million) came from the biotechnology sector, and then outlined the importance of agrifood and bioexports to the province. He indicated that over 30 percent of corn and soybean crops in Quebec were GM and reminded the committee of the extent to which the industry is concentrated and markets integrated within North America. Finally, he raised the issue of the added costs to consumers and producers should labeling require segregation in the processing and transportation of products, in addition to the costs to government of enforcement. The message of potentially lost jobs, investment, and increased costs to farmers and consumers, if too burdensome regulations were imposed on the industry, was clear. Despite his efforts, however, Quebec politicians continue to feel the pressure to promise mandatory labeling. Moreover, experts continue to disagree on the costs, which have now been shown to be substantially lower than the industry claimed (Cloutier 2006; Lalonde 2007). Greenpeace Canada reminded politicians of their promises by dropping five tons of GM corn on the steps of the Quebec National Assembly (Greenpeace Canada 2007). Canada's position on the labeling issue at the Codex will not likely change unless these struggles at the national and regional levels change the national regulatory regime on labeling. Moreover, it is clear that the whole construction of the limited regulation of GM products, based on the substantial equivalence argument, would then be called into question, an aspect that clearly explains a part of the resistance. The transparency argument for consumers to be informed about both the content and the process by which their food is produced has to date not proven sufficient to create that change. Nonetheless, in both Canada and the United States, the issue will not go away even with the power that corporate interests have.

Despite the continued success of the industry at resisting mandatory labeling in Canada and the United States, the picture at the global level is even more mixed and suggests that corporate power alone has not determined the emerging regime of global food governance. The following conclusion reflects on the case study of GM food labeling and what it tells us about corporate power in the food governance process.

Conclusion: Corporate Power, Transparency, and Food Governance

At the global level, biotechnology firms are becoming ever larger and capital more concentrated. The growth of GM crops increases each year as GM technology is pressed on the developing world by bilateral and multilateral agencies such as the World Bank. However, this case study of food labeling suggests that the power of agribusiness, while important, alone does not fully explain the struggle over the international rules affecting the production and exchange of food products. The emergence of consumer, environmental, and other social movements around food issues and labeling, along with divisions among powerful state actors such as the United States and the European Union, have continued to challenge corporate power, both in terms of national regulations and at the Codex. Supporters of mandatory labeling have been able to resist the efforts of some powerful corporate and state actors to stop work on, or discussion of, mandatory labeling. In fact, over the past fourteen years the majority of Codex members have come to favor broader and more mandatory labeling regimes and now implement them at the national or regional level.

The reality is that the structural and instrumental power of corporate actors is uneven and their material interests do differ. The biotechnology industry has not yet gained the foothold in some areas of Europe that it has in the United States and Canada. Food retailers in Europe have viewed their interests in terms of labeling and GM food differently (Falkner, chapter 8, this volume). As the case study indicates, the openness of the Codex process to observers, both corporate and NGO, has also given other interests some voice. The biotechnology industry's voice is heard loudly, but so are other voices albeit with less volume. These other interests have also had an impact in shaping national or regional regulatory regimes. Having faced several food scares, state regulatory failures, and the flawed assurances of experts, EU consumers have become highly skeptical of scientific experts and have demanded stronger precautionary regulation, reflected in the 2003 directive on labeling.

A third element that has challenged the discursive power of corporations has been the norm of transparency, reflected in the consumer desire for more information and labeling of GM foods. This is especially the case in the context of a decline in the trust in governments and corporate actors. One way to restore public trust is, of course, through regulatory agencies that are seen to be transparent and accountable to citizens and at arm's length from corporate interests. The North American trend to deregulate transfers the risk to consumers and producers, especially where the technology continues to raise important questions about long-term safety, traceability, and the environment. Such a trend increases, rather than decreases, the need for information to address the asymmetry of knowledge between the industry and the consumer.

As this case indicates, however, there are also limits on how the discursive power of transparency and consumer demands for a broad right to know, can overcome powerful corporate and government interests. So far the United States, Canada, and the corporate opponents of tighter regulation of GM products have been able to confine the right to know—that is, transparency—narrowly to issues of food safety, based on "sound science" at the Codex and at the WTO. The EU, on the other hand, has chosen to maintain its more precautionary regulations despite losing a WTO dispute. Whether other factors such as ethical or environmental concerns will be seen as legitimate justifications for a broader concept of transparency based on the process of production, not only the end product, remains to be seen and will depend on the power of the actors and interests that demand it. Mounting concerns about the ethical and environmental risks, to say nothing of the safety risks, of global methods of food production and sourcing could be the basis of an emerging challenge to corporate power. Transparency questions thus will remain at all levels of governance in food.

Notes

1. *Traceability* refers to the process of documentation of the production, handling, and transportation of a product, which facilitates the preservation of its identity, in terms of a particular crop or a specific trait. This assures buyers of the quality of the product. In the case of seeds, identity preservation has a long history. While North American biotechnology companies and producers of GM crops complain about how traceability requirements create market barriers due to the costs and difficulties of crop segregation, other sectors, sometimes in partnership with governments, see retail or export opportunities in identity preservation and traceability since they can command a price premium for providing quality assurance to buyers.

2. This bill was introduced in the US Congress in 1999 and again in 2006, but was never passed.

References

Abergel, Elisabeth and Katherine Barrett. 2000. Breeding Familiarity: Environmental Risk Assessment for Genetically Engineered Crops in Canada. *Science and Public Policy* 27 (1): 2–12.

Bernauer, Thomas. 2003. *Genes, Trade and Regulation: The Seeds of Conflict in Food Biotechnology*. Princeton, NJ: Princeton University Press.

Buckingham, Don. 2000. The Labelling of GM Foods: The Link between the Codex and the WTO. *Agbioforum* 3 (4): 209–212.

Buse, Kent, and Kelley Lee. 2005. *Business and Global Health Governance*. Discussion Paper No. 5. London: Centre on Global Change & Health, London School of Hygiene & Tropical Medicine, December. http://www.odi.org.uk/pppg/politics_and_governance/publications/kb_ghg_who_lshtm_series.pdf.

Charman, Karen. 2001. Spinning Science into Gold: Management of Research and Product Development by the Biotechnology Industry. *Sierra Magazine*, July/August. http://www.sierraclub.org/sierra/200107/charman.asp.

Cloutier, Martin. 2006. Ètude écononmique sur les côuts relatifs à l'étiquetage obligatoire des filières génétiquement modifiées (GM) présenté au Ministère de L'Agriculture. October.

Codex Alimentarius Commission (Codex). 2006. *Report of the Thirty-Fourth Session of the Codex Committee on Food Labelling Ottawa*, May 1–6.

Consumers Association of Canada (CAC). 2003. Consumers' Association Withdraws from National Labelling Committee. Press Release, July 28.

Consumers International. 2006. *Report of the Consumers International Delegation at the 34th Meeting of the Codex Committee on Food Labelling.* May 1–5. http://www.consumersinternational.org/Shared_ASP_Files/UploadedFiles/6B24D2A7-BBC3-4CC7-BF29-0302B1B2F7D3_CCFLDay2Report.doc.

Da, Ros, Jerôme and Diahanna Lynch. 2001. Science and Public Participation in Regulating Genetically-Modified Food: French and American Experiences. Paper presented at the European Community Studies Association, Madison, Wisconsin.

DeBoer, Joop. 2003. Sustainable Labelling Schemes: The Logic of Their Claims and Their Function for Stakeholders. *Business Strategy and the Environment* 12 (4): 254–264.

de Clercy, Cristine, Louise Greenberg, Donald Gilchrist, Gregory Marchildon, and Alan McHughen. 2003. *A Survey of the GM Industry in Saskatchewan and Western Canada*. Paper No. 16. Regina: Saskatchewan Institute of Public Policy, May. http://www.uregina.ca/sipp/documents/pdf/PPP%2016_%20GMO.pdf.

Greenpeace Canada. 2007. Where Are Your GMO Labels, Mr. Charest? March 28.

Greenpeace, Option consommateurs, and l'Union des consommateurs 2004. Mandatory Labelling of GMOs Public Opinion Strongly in Support. *Canadian Newswire.* May 10. http://www.greenpeace.org/canada/en/campaigns/ge/latest-developments/mr-charest-where-are-our-labels.

Grocery Manufacturers of America (GMA). 2003. Letter to Christine J. Taylor, Ph.D. US Delegate to the Codex Committee on Food Labelling, U.S. Food and Drug Administration, August 6.

Grocery Manufacturers of America. 2007. GMA Member Brochure. http://www.gmaonline.org/docs/2007/GMAmemberbrochure.pdf.

Hallman, William K. 2000. *Consumer Concerns about Biotechnology: International Perspectives.* Rutgers University: Food Policy Institute.

Harness, Robert L. 1998. Testimony before the Subcommittee on Trade of the House Committee on Ways and Means Hearing on Trade Relations with Europe and the New Transatlantic Economic Partnership. July 28. http://waysandmeans.house.gov/legacy/trade/105cong/7-28-98/7-28harn.htm.

Huller, Thorsten, and Matthias Leonhard Maier. 2006. Fixing the Codex? Global Food Safety Governance under Review. In Christian Joerges and Ernst-Ulrich Petersmann, eds., *Constitutionalism, Multilevel Trade Governance and Social Regulation*, 267–299. Portland, OR: Hart.

International Centre for Trade and Sustainable Development (ICTSD). 2007. Codex: Work on Labelling Biotech Food to Continue. *Bridges,* May 25.

International Life Sciences Institute (ILSI). 2007a. About ILSI. http://www.ilsi.org/AboutILSI/.

International Life Sciences Institute. 2007b. International Food Biotechnology Committee. http://www.ilsi.org/AboutILSI/IFBIC/.

Kuyek, Devlin. 2002. *The Real Board of Directors: The Construction of Bio-Tech Policy in Canada 1980–2002*. Sorrento, BC: The Ram's Horn. http://www.ramshorn.ca/Real_BoD.pdf.

Lalonde, Michelle. 2007. Labelling of Genetically Modified Foods Relatively Inexpensive. *Montreal Gazette,* March 29.

Leiss, William. 2003. The Case for Mandatory Labelling of Genetically- Modified Foods. Paper prepared for the Consumers Association of Canada. November 8. http://www.consumer.ca/pdfs/case_for_labelling_genetically_modified_foods.pdf.

MacKenzie, Anne A. 2001. International Efforts to Label Food Derived through Biotechnology. In Peter Phillips and Robert Wolfe, eds., *Governing Food: Science Safety and Trade,* 49–61. Kingston: Queen's University School of Policy Studies.

National Food Alliance. 1993. *Cracking the Codex: An Analysis of Who Sets World Food Standards.* London: National Food Alliance.

Organization for Economic Cooperation and Development (OECD). 2003. *Agro-food Products and Technical Barriers to Trade: A Survey of the Issues and Concerns Raised in the WTO's TBT Committee.* Paris: OECD.

Pfister, Patrick. 2005. Clashing Arenas or Network Governance: The Challenges of Interplay in GM Food Regulation. Paper presented at the Berlin Conference on the Human Dimensions of Global Environmental Change, Berlin, December.

Prakash, Aseem, and Kelly Kollman. 2003. Biopolitics in the EU and the US: A Race to the Bottom or Convergence to the Top? *International Studies Quarterly* 47 (4): 617–641.

Sell, Susan. 2007. Lobbying Strategies of Multinational Corporations in Biotechnology. Paper presented at the International Studies Association, Chicago, March 4.

Sklair, Leslie. 2002. The Transnational Capitalist Class and Global Politics: Deconstructing the Corporate-State Connection. *International Political Science Review* 23 (2): 159–174.

Smith, Jeffrey. 2003. *Seeds of Deception.* Fairfield, Iowa: Yes Books.

Stewart, Lyle. 2002. Good PR Is Growing. *This Magazine*, May/June. http://www.healthcoalition.ca/goodprisgrowing.html.

Suppan, Steve. 2006. Codex Standards and Consumer Rights. *Consumer Policy Review* 16 (1): 5–13.

Thomas, Courtney. 2006. *The Impacts and Implications of Post-1995 Linkages between the Codex Alimentarius Commission and the World Trade Organization.* MA thesis, Virginia Polytechnic.

US Public Interest Research Group (US PIRG). 2005. At a Standstill: The United States Role in Stalling International Efforts to Label Genetically Engineered Foods. www.uspirg.org.

Veggeland, Frode, and Svein Ole Borgen. 2005. Negotiating International Food Standards: The World Trade Organization's Impact on the Codex Alimentarius Commission. *Governance* 18 (4): 675–708.

Victor, David G. 1997. *Effective Multilateral Regulation of Industrial Activity.* PhD dissertation, MIT.

World Health Organization (WHO) and Food and Agriculture Organization (FAO). 2002. *Report of the Evaluation of the Codex Alimentarius and Other FAO and WHO Food Standards Work.* November 15. Rome: FAO.

World Health Organization (WHO) and Food and Agriculture Organization (FAO). 2005. *Understanding the Codex Alimentarius.* Rome: FAO.

5

Corporate Interests in US Food Aid Policy: Global Implications of Resistance to Reform

Jennifer Clapp

Food aid has long been highly political. It has also long been associated with economic interests, particularly in donor countries. This is especially the case with food aid that is donated in its commodity form, which is typically tied to commodities grown in the donor country. Tied, in-kind food aid has historically been used as a surplus disposal mechanism. While today there are fewer food surpluses to be disposed of, the economic interests involved in the provision of tied food aid in the United States have fought hard to maintain that system. A constellation of actors that benefit from the present system, including corporate lobby groups that represent the grain industries, the milling industry, the shipping industry, and a food aid lobby group representing the nongovernmental groups that distribute the aid, have in effect created a powerful force working to keep the system as it is. But while the United States has kept its in-kind food aid programs intact, there has been growing recognition in both the academic and policy communities that in-kind food aid is problematic. It can distort markets, can depress production incentives in recipient countries, and is also highly inefficient.

As other countries, including the European Union (EU), Canada, and Australia, have made significant steps to untie their food aid in recent years, they have put pressure on the United States to move away from in-kind food aid. This push has come via the World Trade Organization (WTO) negotiations on agriculture as part of the Doha Round. The EU has taken a key role in arguing for a primarily cash-based food aid system on the grounds that the US in-kind food aid programs are distorting trade, both in recipient countries and for other countries selling those commodities. At the same time, the political process around the 2008 Farm Bill in the United States has also stirred calls from within the country, including from US President George W. Bush, to at least partially untie US food aid in order to improve efficiencies and reach more people in need. These developments have sparked enormous debate within the United States over its food aid policy. How this debate plays out is extremely important on an international scale, because the United States is by far the world's largest donor of food aid.

Corporate actors have taken a prominent role in debates over food aid reform in the United States in recent years. This chapter examines how corporate actors influence food aid governance in the United States, which in turn has global significance because of the importance of the United States as the world's largest donor of food aid. I argue that corporate concentration and the historic importance of the grain, milling, and shipping industries to the US economy give these actors a degree of structural power in setting the agenda for discussing food aid reforms. These same industries have also actively lobbied the US government, which has given them a degree of instrumental power. Finally, they attempt to influence public perceptions of the role of food aid policies by engaging in public debate over questions of hunger and food security and its link to food aid.

The examination of this corporate influence reveals two further points. First, corporate actors have a complex relationship with other nonstate actors that also have a stake in food aid policy. A number of key nongovernmental organizations (NGOs), as part of the umbrella group the Coalition for Food Aid, have historically supported a similar position to grain and shipping industry groups because of their own economic interests in the current food aid policy. This position on the part of some NGOs has lent a degree of legitimacy to corporate opposition to food aid policy reform. This tripartite alignment of interests (grain industries, shipping industry, and food aid NGOs) has been labeled by Barrett and Maxwell (2005) as the "iron triangle." But a growing number of NGOs are increasingly moving away from the corporate position and toward a more reformist position. Second, the institutional context within which US food policy is made is particularly susceptible to the influence of private actors, including both corporations and NGOs. Food aid policy in the United States is governed via the farm bills and annual presidential budget requests, which means that the US Congress must approve changes in the policy. This has enabled private actors to engage in public debates over food aid, and to penetrate the policy process via intense lobbying and threats to withdraw political support should the policy change.

The debate over food aid reform in the United States highlights the contested nature of this form of aid and reveals that there are different interpretations of its implications for socioeconomic and environmental sustainability goals. On one side, industry and certain NGOs argue that in-kind food aid is a necessary policy because it uniquely meets the multiple objectives of providing food security in poor countries while at the same time providing benefits for the United States in the form of farm support, export promotion, and national security enhancement. Others, however, including some NGOs that have broken rank, argue that in-kind aid has serious flaws that compromise sustainability objectives. In particular, it is seen to weaken farmer incentives in developing countries by displacing local production and threatening rural livelihoods. It also is seen to be grossly inefficient at the same time that it can harm local environments by introducing genetically modified organisms in countries where they do not have regulatory approval.

After providing a brief history of international food aid trends, the chapter proceeds with a discussion of the various facets of corporate power and its impact on current international food aid governance and policy. The final section examines the impact of in-kind aid for socioeconomic and environmental sustainability by looking at the arguments made by those who oppose in-kind food aid.

Past and Present Trends in International Food Aid

International food aid was first institutionalized as a form of development and humanitarian assistance by the United States, through its 1954 US Agricultural Trade Development and Assistance Act, otherwise known as PL 480, and more commonly referred to as "Food for Peace" (USAID 1995) as part of the 1954 Farm Bill. PL 480 was adopted following the collapse of agricultural commodity prices in the United States in the early 1950s, which was a result of massive grain surpluses. The grain surpluses were partly a product of a more industrialized agriculture, and partly a product of US policy to support its farm community via subsidies and payments following the Second World War (Friedmann 1993). Grain surpluses that resulted from these policies were purchased by the government as a "buyer of last resort," and needed to be disposed of somehow. Distributing them as food aid seemed to be an obvious answer to the problem. Removing surplus grain from the domestic market would help to keep prices buoyant by restricting supply. And shipping it to poor countries would serve a humanitarian goal of feeding the hungry. Food aid in the 1950s and 1960s, then, was very important for the maintenance of a nationally managed agricultural system, reflecting a mercantile approach to agricultural policy on the part of the United States in this period (Friedmann 1993; Cohn 1990). The obvious political benefit of this policy was that it was seen that food assistance could help the US foreign policy goal of supplying food to allies in the developing world during the Cold War (see Uvin 1992; Hopkins 1993).

The United States historically both sold food aid (at a discount) and gave it away in grant form (for those countries with more need, and for emergencies) out of government-held surpluses. These various types of food aid were organized under three Titles of PL 480. Title I constituted concessional sales of food. Title II was grant food aid for humanitarian and emergency uses. And Title III was government-to-government grant aid for development activities. Food aid sales were still considered aid because they were offered at a discount price on easy credit terms, and were aimed at increasing commercial exports once recipients were able to purchase the food at market prices. In the 1950s and 1960s, the United States sold the bulk of its food aid as discount sales (around 60 percent of food aid) with the other forms of food aid making up the rest. All US food aid was explicitly tied to the purchase and donation of US-grown grain, a significant portion of which was to be shipped on US-flagged ships. This latter criterion was designed as a support for the US Merchant Marine, and was seen as being supportive of national security objectives. Following the US lead with respect to the establishment of a food aid program, other donor countries followed suit. They were less overtly political in terms of using food aid as a tool in the Cold War, but most did explicitly tie their food aid donations to food sourced from the donor country. The European Community, Canada, and Australia, for example, all brought in food aid programs tied to production in those countries in the 1960s (Uvin 1992).

At the international level, there was a move in the 1960s to provide a multilateral institutional structure and rules for food aid. In 1963 the World Food Program (WFP) was established as a UN agency to coordinate donations of food aid, in particular in cases of emergencies. Based in Rome, the WFP accepts donations from donor countries and provides emergency relief. Initially assistance from the WFP constituted a small percentage of total food aid delivery, but in recent years it now administers a significant percentage, accounting for some 50 percent of global food aid delivery (IGC 2004). The Food Aid Convention (FAC) came into existence in 1967 as part of the Kennedy Round of the General Agreement on Tariffs and Trade (GATT). The FAC was a partner agreement to the wheat trade convention, and both were negotiated under the International Grains Arrangement during the Kennedy Round. The FAC is a treaty among food aid donor countries that sets out minimum commitments for these countries with respect to food aid donations. The FAC has been regularly updated and renegotiated on a periodic basis by the Food Aid Committee, which meets regularly under the auspices of the International Grains Council (IGC). By the late 1960s, a large food aid industry and an organizational structure to govern it had emerged. At that time, some 20 percent of all official development assistance from rich to poor countries was in the form of food aid (OECD 2005, 11).

Since its early days, food aid has seen some important changes, which reflect changes in the broader global political economy and the food economy in particular (Friedmann 1993). Of particular importance is that rather than being a form of government-held surplus disposal, most food aid today is sourced on open markets (Barrett and Maxwell 2005, 88–89). This shift was in large part a result of restructuring in the global food regime away from government purchase of surpluses, especially after the food crisis of the mid-1970s, which saw sharp increases in food prices. With this shift, the significance of the US food aid program for the broader food and agricultural policy of the United States had diminished. This shift away from food aid as a surplus disposal mechanism has been mirrored in other countries but has manifested differently, in the form of untied food aid. The European Union since 1996 has made a significant shift toward providing primarily cash-based food aid for local and regional purchases in developing countries (OECD 2005, 21–22; Clapp 2008). Australia and Canada have also shifted away from policies that tie their food aid to donor-country commodities and toward funds for purchasing the food in or near the recipient country. Other countries have now begun to follow the EU lead. In 2004 Australia moved to untie up to two-thirds of its food aid. Canada untied up to 50 percent of its food aid in 2005 (CIDA 2005) and adopted a policy of 100 percent untied aid in 2008.

Another major difference between the early period and today is the significance of food aid in terms of overall assistance. Food aid as a proportion of overseas aid has declined markedly, to around 5 percent of official development assistance today (OECD 2005, 11). The proportion of food aid that goes to emergencies has also risen sharply. In the 1960s emergency food aid made up around 20 percent of food aid donations, while in 2005 it accounted for 64 percent of food aid (FAO 2007, 47). In line with this trend, the US sale of discounted food as aid has also declined considerably.

A further new development in US food aid policy was the introduction in the 1990s of the practice of "monetization." The emergence of this practice is linked to a growing reliance on NGOs to act as distributors of food aid in the recipient countries, rather than government distribution. This practice, as discussed below, has been criticized for the role it can play in distorting markets in recipient countries. Food aid–delivery NGOs have been encouraged to sell (or monetize) a portion of the food aid they are given to distribute on local markets in the recipient country as a means by which to raise funds to cover the cost of food distribution, and to fund their other development projects (Barrett and Maxwell 2005; Oxfam International 2005, 23). This practice saves donor countries money in terms of allowing the in-kind aid to pay for its own distribution by selling a portion of it in developing countries. Today, around half of all in-kind food aid donations are monetized in this way, though for some NGOs, that figure is much higher (OECD 2005, 12).

Although some of the key features of the food aid policy landscape are very different today than was the case in the early period, the United States still remains the world's largest food aid donor. Today it provides some 50–60 percent of all food aid globally, which makes US policy and its impact especially important in terms of global food aid trends and governance. US food aid donations are also still heavily tied to US producers, processors, and shippers, with 100 percent of food aid donations from the United States to be supplied in-kind, in the form of food. Further, 50–75 percent of that aid, as spelled out in the US Cargo Preference regulations, must be shipped on US-flagged ships. Additionally, half of all US food aid is also required to be processed and bagged in the United States (OECD 2005, 21). These requirements have been in place in more or less the same form since the early days of US food aid. The fact that food aid is now sourced on open markets rather than from government-held surplus stocks, however, has meant that the stake that the grain companies and the millers have in the policy has become even more entrenched.

Since the 1970s, numerous studies have examined the impact of in-kind food aid in both local and global contexts. Within this literature, there has been a growing academic consensus that in-kind food aid can have market-distorting effects in recipient countries. In-kind food aid is also critiqued for its inefficiencies that result in less food being provided than would be the case if cash were provided to purchase food either locally in recipient countries or regionally in developing countries (for an overview, see OECD 2005; Barrett and Maxwell 2005).

Although food aid today accounts for only around 2 percent of the global grain trade, it can result in serious distortions to local economies in developing countries where it is delivered. In-kind food aid tends to be pro-cyclical—that is, the amount of food aid given varies with supply—and this makes it somewhat unpredictable for recipients. When food stocks are high and prices are low, more food aid is donated; when stocks are low and prices high (just when poor food-deficit countries need aid the most), food aid donations tend to fall. These fluctuating amounts of food aid can be very disruptive to local economies. And because it typically takes at least four to five months for a food aid shipment to get to where it is headed, it often does not arrive in a timely manner. In addition, the monetization of food aid means that instead of targeting food aid to those most in need (i.e., those who cannot afford to buy it), it is sold on local markets. Again, this leads to displacement of commercial imports, and also depresses demand for locally produced food. At the same time, there is no guarantee that the food is actually reaching those who are most in need of it.

On top of distorting markets, in-kind food aid is typically much more expensive to provide, which in the end means less food being provided to those in need. Often there is food in the local market available at prices much lower than the cost of shipping aid halfway around the world, but people cannot afford to buy it. It is also difficult to take recipient preferences adequately into account with in-kind food aid. There are numerous instances where countries were given culturally inappropriate foods, or where food aid was sent that that did not meet health, safety, or environmental regulations of the recipient country.

The specific critiques of in-kind food aid have been discussed in the academic literature for decades, but policy has been very slow to adapt. As mentioned above, some change is evident among some donors. But the United States, as the largest food aid donor, has not altered its policy, despite increasing pressure to do so by other donors as well as by academics and some policymakers within the United States. This continuation of an in-kind food aid policy has been at the center of a number of food aid controversies in recent years.

In 2002 the United States sent significant quantities of food aid to Southern Africa in response to food shortages and drought. The aid was sent in the form of whole-kernel maize, and it was found to contain genetically modified organisms (GMOs) that were not approved for consumption in those countries. The recipients had not been notified of before the shipments were sent. A number of countries in the region initially refused to accept the genetically modified (GM) food aid, partly as a health precaution, and partly on the grounds that it could contaminate their own crops, thus hurting potential exports to Europe in the future. The fact that the aid was provided in-kind from the United States meant that food preferences and environmental requirements of the recipients could not be easily accommodated. The incident sparked a major international controversy, with the United States taking a defensive position on the issue, claiming that it was impossible to source non-GM crops from the United States. It refused to give cash instead of in-kind aid, on the grounds that it has traditionally given in-kind aid, and has done so for fifty years (Clapp 2005).

Just when the controversy over GM food aid was waning, in-kind food aid again became a hot political topic with the heating up of the WTO Doha Round of trade talks in 2003 and 2004. In the context of discussion over reduction of agricultural subsidies in industrialized countries, the European Union insisted that the way countries provide food aid be considered for reform in the trade talks because it sees in-kind food aid as an unfair and market-distorting export subsidy to producers in donor countries. The EU has been targeted to reduce its own export subsidies as part of the trade deal. In response, the EU has called for "parallelism" in terms of what are considered export subsidies, and has said that it will not commit to reductions in its own export subsidy programs unless the United States undertakes reforms to its in-kind food aid (Clapp 2004). This insistence on including food aid as a trade issue in the WTO talks has sparked heated debate in the United States, which stressed in official talks that in-kind food aid does not have the same market-distorting impact as other forms of export subsidy. Despite the basic disagreement between the United States and the European Union on this matter, WTO members agreed in principle in 2004 that some disciplines should be placed on in-kind food aid so as to ensure that it is not market-distorting (WTO 2004).

The European Union pushed the issue at the WTO Ministerial Meeting in Hong Kong in December 2005. The ministerial declaration from this meeting announced that the issue would be resolved in the current round, with a commitment to end commercial displacement. Disciplines would be placed on food aid, in particular to disallow in-kind food and monetization except in specific circumstances where it could be guaranteed not to be market distorting (WTO 2005). To address emergency situations where it is deemed by the international community that in-kind food aid is an appropriate response, the proposal calls for a "safe box" that would exempt in-kind aid from disciplinary action in those instances. Food aid in the form of cash would also be placed in the "safe box," and thus would be exempt from the disciplines. This would ensure that food aid itself is not the issue. Rather, the concern is whether food aid distorts markets and constitutes an agricultural subsidy for the donor country. The details of this arrangement have yet to be worked out in the context of the Doha Round. It is still unclear who determines what constitutes an emergency situation and whether in-kind aid in such cases is likely to distort markets.

At the same time that food aid became an issue in the WTO trade talks, the Bush administration, at the urging of the main US aid agency—the US Agency for International Development (USAID)—pushed for allowing up to one-quarter of its food aid budget in the form of cash to enable local and regional purchases of food in developing countries. US President Bush even went so far as to put a request to this effect in his budget proposal for both 2006 and 2007, mainly on the grounds that it would enable US aid to reach more people in a more timely fashion. The president's budget proposal for 2006 asked for $1.2 billion to be allocated for food aid programs, but it also asked that one-quarter of that amount, $300 million, be given as cash assistance, rather than as in-kind donations. The 2007 budget request asked for no funds at all to be allocated to food aid sales (known as Title 1) (Monke et al. 2006). But the proposals were halted both years by the US Congress, which faced strong resistance from powerful lobby groups representing the agrifood industry, shipping companies, and certain key private voluntary organizations that have a stake in the current organization of US food aid.

The timing of the US president's budget requests coincided with the work of the US Congress in drafting the 2008 Farm Bill, which is where PL 480 resides and is reauthorized or amended each time a farm bill passes. President Bush put forward his proposal for the farm bill, which included allowing up to 25 percent of the US food aid budget to be in the form of cash for local and regional purchases in developing countries to increase flexibility in response to food crises. In mid-2008 the farm bill was finally adopted, and included only a small pilot project for local and regional purchases (see Partnership to Cut Poverty and Hunger in Africa 2008). Agribusiness, the shipping industry, and a number of NGOs were especially active in their attempts to influence US food aid policy in these various contexts. The discussion below focuses mainly on the activities of the two corporate sides of this triangle, the shipping industry and the agrifood companies, and their complex relationship with the NGOs that also have a stake in the food aid issue.

Corporate and Private Interest Power in US Food Aid Policy

Corporate actors have taken a keen interest in attempting to influence US food aid policy through an overall strategy of resisting reform to current policy and practices. Many have attributed the slow pace of change on this policy in the United States compared to other countries to the strength of the collection of lobbyists that have fought hard against reform. The focus here is on the various ways corporate players have attempted to influence the food aid policy process in the United States through their use of structural, instrumental, and discursive power. These various forms of corporate power overlap in many ways, and while each is distinct, each also feeds into and plays off of the others.

Structural Power of Corporations with Food Aid Interests

Because of their importance in the provision of food aid, corporate actors have been able to exert a degree of structural power over food aid governance in the United States. This influence has emerged as a result of several factors, including the concentration in the grain, milling, and shipping industries, which gives them particular weight in political decision making with respect to setting the agenda for food aid policy discussions. While instrumental power in terms of lobbying policymakers is also important here (as discussed below), it does appear as though some parts of the US government, such as the Department of Agriculture and Congress, are not just responding to lobbyists' pressure, but also are defending the interests of these industries because of their importance to the US economy more broadly. At the same time, food aid is clearly very important to these industries, giving them a strong incentive to highlight the significance of their role in the economy as a way to preempt potential policy changes that would harm their interests. This strength is asserted in terms of broader threats that if proposed reforms to food aid were to be implemented, that broad support for food aid from industry would be weakened considerably.

There is significant concentration in the global grain industry, which is dominated by a handful of global corporations. Today there are just four main grain trading companies that make up the bulk of the global trade in grains: Cargill, Archer Daniels Midland (ADM), Louis Dreyfus, and Bunge (ActionAid International 2005, 12). These companies, which are all privately owned, are truly global operations, and their share of the market is significant (Murphy 2006, 9–10). Cargill, for example, which operates in over 160 countries around the world, had sales and revenues in 2005 of over US$70 billion (Murphy 2006, 9; Murphy and McAfee 2005, 6). Cargill and ADM together account for 40 percent of all US grain exports (Murphy 2006, 14).

Because of their importance in the US and in global grain markets, the main grain trading companies are key players in the current food aid operations of the United States. The fact that US food aid is now purchased on open markets translates into more direct benefits for grain handling and processing companies (Barrett and Maxwell 2005, 89). Because of the US rule that US food aid be sourced domestically, there are a limited number of potential providers. According to Dugger (2007b), more than half of US food aid in the Food for Peace program over the 2004–2007 period was purchased from just four large transnational agrifood companies and their subsidiaries (ADM, Cargill, Bunge and Cal Western Packaging). Cargill alone is reported to have sold $1.09 billion in grain to the US government for food aid between 1995 and 2005 (Thurow and Kilman 2005). The concentration in the grain industry means that there are often just a few companies bidding on contracts, and thus their prices are often above market rates due to a lack of competition. This concentration, combined with inefficiencies in the procurement process in the United States, translates into a significant price premium for food aid over going market rates. According to Barrett and Maxwell, who calculated these premiums, the extra amount paid to grain companies for food aid over going market prices ranges from around 3.2 percent for wheat, to more than 70 percent for corn. On average, US-supplied food aid costs the government 11 percent more than market prices (Barrett and Maxwell 2005, 91–92).

It is not just the grain industry that reaps extra profits from US food aid programs. The shipping industry earned some US$1.3 billion over the 2003–2006 period in contracts to move food aid to its destinations (Dugger 2007a). And just five shipping firms were awarded contracts for over half of the US$300 million spent on shipping US food aid in 2004 (Dugger 2005). Because of the rule that US in-kind food aid must be shipped on US-flagged ships, relatively few shipping lines are qualified to bid for contracts to carry food aid. As Barrett and Maxwell (2005, 94) point out, "Only 18 shipping companies were qualified to bid on food aid contracts in the 1990s," a number that had dropped to just 13 by 2002. The concentration is even more pronounced for the "freight forwarders," who, as Barrett and Maxwell explain, are firms that coordinate international transportation of the aid for the NGOs that are engaged to distribute the food aid in developing countries (Barrett and Maxwell 2005, 94). This concentration at the shipping and freight end of food aid has led to very high costs for transportation of food aid, which now constitute more than half the cost of providing in-kind food aid (GAO 2007, 1).

For some individual firms involved in food aid provision, the contracts they win are essential for their survival. For example, food aid made up some 23 percent of all hard white wheat exports for 2001–2002, and 17 percent of hard red wheat exports in 2002–2003. According to the wheat industry, losing these food aid contracts would be the equivalent to losing the Japanese market for wheat (Pratt 2006). The US Wheat Associates (2007), a lobby group for the wheat industry, stated that "losing in-kind donations and monetization would be a difficult blow for the U.S. wheat industry." Food aid also accounts for around 10–20 percent of all US rice exports (information presented by Oxfam, in US Committee on Agriculture, House of Representatives 2005, 58). While food aid contracts are not essential for the viability of the US shipping industry as a whole, a relatively few freight lines do profit enormously from it. Food aid is vital for the survival of a few key shipping lines, such as Waterman Steamship Corporation and Liberty Maritime, which rely heavily on food aid contracts as a main source of income (Barrett and Maxwell 2005, 96; Lowenberg 2008).

The grain industry in particular has used its structural power within the marketplace to threaten to withdraw support for US food aid programs if the government allows a portion of its aid to be in the form of cash for local purchases. These threats have stressed the importance of grain sales to the US economy. According to industry representatives, US$1 billion in grain sales translates into US$2.7 billion worth of domestic economic activity (cited in Hedges 2007). The US$1–2 billion spent on the purchase of US grain for food aid programs thus has a significant economic impact within the United States, a factor that lends structural power to the grain industry over decision making and agenda setting with respect to food aid policy.

Instrumental Power in Food Aid Policy

Corporate actors have been especially active in lobbying the US government regarding food aid policy. This lobbying has taken place both in the context of the WTO negotiations on food aid and in President Bush's proposed policy shift to allow a portion of food aid to be given in the form of cash for local purchases in developing countries. Bush's proposal was made as a request for the farm bill process as well as in his budget requests in recent years. Congressional approval is required for changes to food aid policy in the context of both the farm bills and the US president's budget. Thus lobbying efforts have been targeted at the US Congress in particular, which has been especially susceptible to lobbyists' influence. Because the reform proposals emanating from the WTO as well as from the US president incorporate the notion of reducing the amount of in-kind food aid given by the United States, corporate players have been especially active in lobbying against this idea. They have operated through a number of industry associations and lobby groups that represent various agribusiness interests, including the wheat, rice, milling, and shipping industries. The wheat and rice industry lobby associations often portray their interests as congruent with those of farmers. It should be noted that this is part of their discursive strategy (discussed below) to represent themselves as speaking for the small producer. Most of the members of these industry associations, however, are large-scale commercial operators.

US food aid programs were the subject of a number of hearings by several congressional committees in the 2003–2007 period, because of the importance of these programs in the ongoing WTO negotiations, because of the administration's proposal to reform food aid, and more recently because of their relevance for the farm bill. Corporate actors, along with food aid–delivery NGOs and government officials, lobbied Congress extensively and have been among those called as witnesses. The wheat industry has been especially active in lobbying on this issue (e.g., WETEC, USWA, and NAWG 2001, 2003). Many of the witnesses called in the congressional hearings dealing with the topic have used the opportunity to voice their strong opposition to the reform of current US food aid policies. Industry associations, representing the agrifood industry as well as the shipping industry, have also issued a series of press releases and issue papers on food aid over the last few years, in an attempt to influence members of Congress who would be voting on the various pieces of legislation in which changes to food aid policy are or might be included.

In June 2005, for example, a hearing before the Committee on Agriculture of the US House of Representatives heard from industry association representatives from the wheat, rice, soy, and milling industries, in addition to a number of NGOs that deliver food aid and NGOs that work on policy issues (US Committee on Agriculture, House of Representatives 2005). At this hearing, those testifying on behalf of industry were unanimous in expressing their desire that the United States not give way to the EU proposal for cash-based food aid at the WTO, and advised the US Congress not to accept the proposal of the Bush administration to allow for a portion of US food aid to be given as cash. From the standpoint of the grain industry, only US-grown commodities should continue to be given as US food aid.

One of the tactics used by the corporate lobby is to threaten to withdraw their political support for food aid if the proposed reforms are undertaken. John Lestingi, representing the US Rice Producers Association and the USA Rice Federation, stated that "we see no benefit toward strengthening global food security from either proposal, and any mandated requirement that food aid be provided exclusively as cash would undercut the strong support within agriculture for food aid" (quoted in US Committee on Agriculture, House of Representatives 2005, 53). Barbara Spangler, from the Wheat Export Trade Education Committee (WETEC), stated in her testimony to Congress: "Do not expect our growers to be calling on you to support the cash-only programs" (quoted in US Committee on Agriculture, House of Representatives 2005, 31).

Threats such as these have filtered through to Congress, and appear to have had an impact. According to Democratic Representative Tom Lantos, who at the time chaired the House Foreign Relations Committee, separating food aid from the US shipping and agrifood industry would be "a mistake of gigantic proportions . . . because support for such a program will vanish overnight, overnight" (quoted in Dugger 2007a).

And the chair of the House Committee on Agriculture, Bob Goodlatte, when USAID presented its argument for shifting $300 million to cash aid, made it clear that the proposal would undercut the support of the constituents of the committee. He asked: "Is the administration aware of the risk that it takes in proposing something like this and undercutting the base of support for these food aid programs?" (quoted in US Committee on Agriculture, House of Representatives 2005, 16).

The strength of the agribusiness lobby in convincing the US Congress not to endorse any proposal in or outside of the United States with respect to reform of food aid governance has been formidable. Some influential senators initially endorsed the proposal of the Bush administration to allow some cash assistance as food aid, but then backed down under corporate lobby pressure (Haider 2006). Previous administrations had not dared to confront this powerful lobby with respect to in-kind food aid, the reason being the extent of the influence of this lobby over members of Congress (Timmer 2005, 1). In a press release following the 2005 hearing, Bob Goodlatte, Chair of the House Agriculture Committee stressed that "I believe this change carries the great risk of undercutting Congressional support for food aid programs and thereby the ability to provide aid to those in need" (quoted in US House of Representatives Committee on Agriculture 2005b). Subcommittee Chair William Jenkins acknowledged the importance of the industry lobby when he remarked that "it is important to understand that food aid is important to American farmers who produce food, American businesses who process, package and transport the food, and the American private voluntary organizations who are on the ground making sure the food goes to those who really need it" (quoted in US House of Representatives Committee on Agriculture 2005b).

The shipping industry, another especially strong lobby group, has also been active in trying to convince the US Congress to keep present food aid policies intact (American Maritime Officer 2006). In a television interview in 2005, Republican Senator Charles Grassley lamented that "it's too bad we can't reform Cargo Preference but the maritime union and industry is so powerful in Congress that I usually get about 40 votes for doing that. I've tried several times" (quoted in Nichols 2005).

Adding to the force of the industry arguments is the fact that a number of food aid–delivery NGOs as part of the Coalition for Food Aid, an umbrella group representing over fifteen food aid NGOs from 1985 to

2006, including World Vision, Save the Children, and CARE, have also lobbied the US Congress.[1] This NGO lobby made arguments that are somewhat similar to that of the grain and shipping industries with respect to the cash versus in-kind issue. Many of these NGOs have a strong economic interest not only in maintaining current levels of US food aid, but also the current system of in-kind aid. According to Barrett and Maxwell, the main NGOs that deliver food aid rely on that activity for a significant portion of their revenues. Moreover, monetization of the food aid helps these organizations raise funds for other non-food-related projects that might not otherwise attract funds (Barrett and Maxwell 2005, 98–100). Because of their reliance on food aid as a key source of funding, the threats of industry to withdraw support for food aid programs should the United States reform its policy have been especially worrisome for these groups (Thurow and Kilman 2005). The Coalition for Food Aid hired a full-time lobbyist to convince the US Congress to maintain in-kind food aid levels. This group was initially skeptical of the proposals for a portion of US food aid to be given as cash. Because of the similarity of this group's position to that of the industry groups, the food aid NGOs gave a degree of legitimacy to the position of the industry groups, and together they formed a powerful bloc fighting against food aid reform.

In 2006, the strength of this bloc began to weaken. There was a growing distance between the positions of some of the food aid NGOs and industry on the one hand, as well as a split among the food aid NGOs on the other hand. In the face of arguments put forward by the US administration, academics, and food security advocacy groups regarding the problems associated with in-kind food aid, several food aid NGOs—including CARE, Catholic Relief Services, and Save the Children—began to distance themselves from the Coalition for Food Aid. These organizations had previously defended the in-kind food aid programs and the practice of monetization. But they now began to endorse the idea of local and regional purchases in cases of emergencies, effectively lending support to the proposal to allow for some cash aid (see, for example, CARE USA 2006; CRS 2007). As a result of this move on the part of some of the larger food aid–delivery NGOs, the Coalition for Food Aid dissolved in April 2006. It reemerged a few months later under the name of the Alliance for Food Aid, which represents those NGOs that did not take as strong a stand as these groups with respect to cash aid. But while the Alliance for Food Aid has been cautious with respect to monetization and cash aid, it has endorsed the idea of a pilot program for local and regional purchases, provided this is carried out with additional funds rather than coming from the existing food aid allocation (Levinson 2007).

Although there has been a change of position among the food aid NGOs with respect to how to handle the cash versus in-kind issue, corporate lobby groups have continued with their firm opposition to any local or regional purchases. Rather than downplaying the need for rapid assistance in the case of emergencies, corporate groups have instead begun to argue for prepositioning of US food aid in foreign locations as a means to readily supply food in emergency cases. A new coalition of industry groups, known as the Agricultural Food Aid Coalition, consisting of the major agricultural industry lobby associations, promoted this position in US congressional hearings on the topic in early 2007 (Agricultural Food Aid Coalition 2007). Key members of this coalition have also participated in an Ad Hoc Coalition in Support of Sustained Funding for Food Aid, along with maritime groups representing the shipping lobby, which has also called for prepositioning of food aid (Ad Hoc Coalition in Support of Sustained Funding for Food Aid 2007).

Corporate players also have been able to exercise instrumental power at the international level with respect to food aid policy. The Food Aid Committee, which negotiates and administers the Food Aid Convention, is itself housed in the International Grains Council. The IGC is primarily an organization for grain trading countries, but its annual meeting is heavily dominated by industry representatives. Unlike UN-based organizations where annual meetings and treaty renegotiation sessions are open to observers, the IGC meetings are effectively closed to the public, with enormous registration fees for participants, likely geared toward the grain companies. Because the Food Aid Committee meetings that are held alongside the IGC gatherings are also effectively closed to observers, it is difficult to know the extent to which this industry influence at the IGC filters through to negotiations over the Food Aid Convention.

Discursive Power of Corporate Players in the Food Aid Debate

Corporate actors have actively joined the public debate over food aid in an attempt to shape perceptions on the role of in-kind aid in providing food security for poor countries as well as benefits for the United States. The individual companies that win most of the food aid contracts, such as Cargill and ADM, have not themselves stepped into this public debate, but rather the industry associations representing the grain and shipping industries have taken the lead. The US Wheat Associates, for example, launched a campaign in 2006 to "Keep the Food in Food Aid," promoted on its website, while other grain associations have also stressed the need to keep current food aid programs intact (see, for example, WETEC, USWA, and NAWG 2006). Press releases, interviews given to the media, publications, and testimonies in congressional hearings are prominently posted on their websites. Through these public statements, these actors have consistently promoted a number of key arguments with respect to the need to preserve the current structure of US food aid policy. By repeatedly making these arguments, these actors have influenced the public debate on the topic. Thus far the corporate version of the story has won the support of members of Congress. But other government officials, such as the head of USAID and the US president, as well as a number of critics, are not convinced, as will be explained in the following section. Below are the main arguments geared to the US public that are typically presented by industry players with respect to this debate.

Perhaps the most prominent industry argument is that in-kind food aid is a source of pride for Americans. For example, John Lestingi, speaking on behalf of the rice industry, stresses this point by saying that a move to allow some cash-based assistance "would deprive the U.S. agricultural community of their sense of pride and compassion" (quoted in Thurow and Kilman 2005). Similarly, Bart Ruth, from the American Soybean Association, stated that "anything that would restrict the generosity of the American public is not something that we can trade away at the WTO" (quoted in US Committee on Agriculture, House of Representatives 2005, 35). A move to a cash-based food aid system would, according to industry groups, benefit "foreign countries" that are competitors, rather than Americans. An editorial in the *Farm Bureau News*, published by the American Farm Bureau Federation, stressed that "for Americans, food aid is about more than international policy, it's part of who we are" (Finnerty 2004).

It is also clear from the engagement of industry associations and the food aid debate that they see in-kind food aid as important for the development of future commercial markets for US grain exports. Barbara Spangler, representing the Wheat Export Trade Education Committee stressed that "we in the wheat industry believe food aid is an investment. We are investing in the recipient country's future, both in the humanitarian and in the business sense, and we are investing in the great regional and global security that accompanies growing prosperity and stronger private industries that are helped by donations of U.S. commodities" (quoted in US Committee on Agriculture, House of Representatives 2005, 30). Representatives from other grain associations have also made this point, by referring to food aid recipients as their "future customers" (quoted in US Committee on Agriculture, House of Representatives 2005, 30). The domestic benefits are also highlighted, with frequent references to benefits that flow to American farmers. While the statements of industry groups often invoke American farmers as key beneficiaries of in-kind aid policies in their appeal to sway public opinion, the members of these associations are typically large-scale commercial operators, rather than small family farmers.

Both the grain and shipping industries have argued that it is also in the national security interest of the United States to keep its food aid in-kind as opposed to a cash-based system. The rationale here is that US-grown grains are essential not only to provide food for the poor who might otherwise be tempted to join groups that may be hostile to the United States, but also that the US label on the bags of grain provides a powerful symbol of American goodwill, promoting a positive attitude toward the United States abroad. The representative from the North American Millers Association, Jim Madich, was explicit about this when he said that with food aid, "We are essentially reducing the pool of hopeless and disenfranchised people from whom our adversaries draw" (quoted in US Committee on Agriculture, House of Representatives 2005, 27). This sentiment has also been picked up by government officials, who have made similar comments. Senator Pat Roberts, for example, stressed that "food aid programs are one of the most important tools we have in the fight against terrorism" (quoted in Hillgren 2006). The shipping industry has stressed the national security aspect of in-kind food aid as well. The cargo preference regulations are in place, the industry argues, largely to ensure the economic viability of the US shipping industry so that American ships are available if necessary in cases of national security emergencies (American Maritime Officer 2006).

Industry groups have further argued that a move to a cash-based food aid policy would fuel corruption, and thus be much less able to meet food security objectives in developing countries. Bart Ruth of the American Soybean Association stressed the industry's opposition to the idea on these grounds: "We are strongly opposed to cash only grants partly because we will be exposing large amounts of cash to potentially corrupt government officials in recipient countries" (quoted in US Committee on Agriculture, House of Representatives 2005, 29). Similarly, Tom Mick of the US Wheat Associates was quoted in the media as saying that "cash has a tendency to disappear in a lot of third world countries. But when you have 35,000 tons of grain coming in, it's kind of hard to steal that" (cited in Pratt 2006).

Some industry representatives have even gone so far as to stress the importance of keeping the current regulations for 100 percent in-kind food aid in the United States on the grounds that US grain is of better quality and more nutritious, making it an important component of policies to address HIV/AIDS in developing countries. Industry groups are also calling for more long-term, nonemergency aid that is required to be bagged and fortified before being shipped, because a stable and reliable source of supplemental food is seen to be crucial in the fight against AIDS (Testimony of Jim Madich, in US Committee on Agriculture, House of Representatives 2005, 27; Testimony of Bart Ruth, in US Committee on Agriculture, House of Representatives 2005, 29).

By appealing to American pride, the importance of developing future markets, protection of national security, and the fight against both corruption and AIDS in the developing world, industry is seeking to frame the debate over food aid on its own terms. Their aim is to develop an argument under which in-kind food aid is seen as the right and moral approach to food aid. The industry version of the benefits of in-kind food aid, however, is up against an equally powerful argument backed by decades of academic research on the economic inefficiencies and political abuses of food aid, as explained below.

Implications for Sustainability

As indicated in the first section of this chapter, the implications of in-kind food aid for the question of sustainability are numerous and wide-ranging, with important economic, social, and environmental consequences. In-kind food aid, as noted above, has been critiqued in the academic literature since at least the 1970s, on a number of grounds: it is inefficient and costly; it distorts local markets in developing countries and distorts global trade in grains; and it makes it harder to accommodate food preferences and environmental concerns of recipients. With corporate players pushing hard to keep the in-kind food aid policy firmly in place in the United States in the face of growing calls for reform both within and outside of the United States, the sustainability implications of corporate power with respect to this issue are indeed significant. A number of recent reports sponsored by international organizations, the US government, and development and agriculture NGOs detail the effects of continued tying of food aid (e.g., OECD 2005; GAO 2007; FAO 2007; Oxfam International 2005; Bread for the World Institute 2006). These reports have outlined a story that is directly counter to most of the arguments put forward by industry players for keeping in-kind aid as detailed in the previous section.

A main critique of in-kind food aid is that it can lead to a reduction in the number of hungry people who receive food aid. In-kind aid is widely seen to be more costly in economic terms than the provision of cash for local or triangular purchases, such that a continuation of in-kind aid ultimately translates into less food disbursed, which can cost lives. There is a wide consensus in the academic and policy community that in most circumstances, cash aid is more effective and efficient (OECD 2005, 37). According to a study by the Organization for Economic Cooperation and Development (OECD), tied food aid costs on average 50 percent more to provide than local purchases in developing countries, and 33 percent more than triangular (third-country) purchases (OECD 2005, 13). As noted earlier, this higher cost is tied both to a lack of competition among grain companies, which has driven up prices, and to the shipping costs to move food aid halfway around the world. In the US case, the higher cost has translated into less food shipped overseas as food aid, with a 52 percent decline in average tonnage of food aid delivered over the 2002–2007 period (GAO 2007, 1). According to a report of the US Government Accountability Office (GAO), around 65 percent of US food aid costs go toward transportation of the food. The GAO estimates that each US$10 per metric ton reduction in the cost of transporting food aid could feed an additional 850,000 people in an average hungry season (GAO 2007, 15–16). These inefficiencies, particularly those tied to higher grain premiums that result from a lack of competition among the large grain companies, need to be kept in mind in the face of industry claims that cash aid facilitates corruption in developing countries.

A further inefficiency of in-kind aid that has been identified is the issue of timing. It takes on average four to six months to deliver US food aid from donor to recipient, due to the numerous rules and procedures that govern in-kind aid (GAO 2007, 17–18). Because of this time lag, in-kind aid cannot be easily relied on to reach emergency crises in a timely fashion, again potentially costing lives especially in emergency situations. Time lags are also problematic in that by the time aid arrives at its destination, it can (and often does) coincide with local harvesttimes, which results in oversupply and declining prices on local markets, which in turn can be harmful to local farmer incentives. In addition, there have been recent reports of food shipments that are spoiled or rotten by the time they arrive in the recipient country (GAO 2007, 32).

Local markets in developing countries can also be distorted by in-kind aid. Because it tends to be pro-cyclical in nature, in-kind aid perpetuates a situation of uncertainty for recipients (OECD 2005 26; Bread for the World Institute 2006, 9).[2] It is difficult for farmers to plan which crops to plant when they do not know how much or what kinds of foreign food will be entering the market. Further, it is extremely difficult for farmers to maintain sustainable livelihoods in situations where cheap imported grains outcompete the locally grown products they are trying to sell on domestic markets. The practice of monetization is also widely seen to be disruptive to local markets in recipient countries, which frustrates efforts to promote local economic development (FAO 2007, 38; Oxfam International 2005, 23). Around 10 percent of US food aid was monetized in 1990, but this figure increased to over 60 percent by 2002 (OECD 2005, 36). Monetization of in-kind food aid typically results in a drop in prices for grains on local markets, again providing disincentives for local producers (FAO 2007, 38). And in-kind aid, especially nonemergency aid, can result in the displacement of commercial imports, which is seen to be distorting to global markets for grains (FAO 2007, 44; Oxfam International 2005, 18).

In-kind food aid has the potential to change food preferences in recipient countries, as was seen in a number of cases where wheat was introduced to areas where it is not normally grown and has resulted in dependencies (OECD 2005, 33). It also can be problematic when foods with undesirable characteristics from the viewpoint of the recipient are shipped against their wishes. The GMO food aid incidents outlined earlier in this chapter highlight his problem, as recipient countries in Southern Africa were concerned about the potential health impacts of accepting GMO maize (Clapp 2005; Zerbe 2004). Food aid containing GMOs has potentially negative environmental and economic implications as well. The key concerns of the Southern African countries with respect to the GMO food aid that they received in 2002–2003 included a worry that some grains might be planted and lead not only to a disruption of the biodiversity in the region, but could also reduce their ability to export grain to the European Union, where those GMOs were not approved for import (Clapp 2005).

These various critiques of in-kind food aid highlight social, economic, and environmental impacts of the US food aid policy. The arguments have proven to be a powerful motivator for food aid policy shifts in the other major donor countries in recent years. Whether these arguments will penetrate the political debates over food aid in the United States with enough force to lead to a policy shift remains to be seen. The power of corporate players—structural, instrumental, and discursive—has thus far enabled their version of in-kind food aid to prevail in the policy process. Part of the reason for this is the institutional setting in the United States, which puts food aid policy in the hands of Congress via the farm bills and via the need for congressional approval for the US president's annual budgets. Corporate power will likely keep the institutional setup the way it is, because any change along these lines will also require congressional approval (see Clapp 2008).

Conclusion

Corporate actors have been important players in the debates over food aid in recent years. Their role and significance is especially evident in the case of the United States, the world's largest donor of food aid. Corporate actors have long benefited economically from the present configuration of US food aid, which is based on in-kind donations and tied to rules and regulations that seek to benefit US firms. Because of these benefits, corporate actors have been strong supporters of these policies. But recent years have seen pressure for reform away from 100 percent in-kind aid from several quarters. The European Union has targeted US in-kind food aid as an unfair trade subsidy and would like to see it prohibited under WTO rules. And within the United States, the USAID has teamed up with the Bush administration to bring forward proposals that would allow 25 percent of US food aid to be cash-based.

These attempts to push for a delinking of US-grown commodities from US food aid policies have not sat well with the corporate actors and some NGOs that benefit from the current system. The grain companies, milling companies, and shipping companies have attempted to exert their influence over the policy process through a number of channels, and have made use of the similarities at times with some NGOs to enhance the legitimacy of their position. Corporate actors have exerted structural power, through their market significance, as well as instrumental power, in the form of corporate lobbying, and discursive power, by engaging in public debate over the issue. These various forms of power in this case have overlapped, making clear distinctions between them somewhat difficult. But together these forms of power have been influential, and combined with the institutional setting that is particularly susceptible to corporate influence, are a main reason why the proposed policy changes on the table have not yet been adopted. This is despite widespread academic and policy studies that demonstrate the significant costs to developing countries of the current system, as well as high-level support from both the US president and the USAID for a policy change.

The sustainability implications of tied food aid policies, particularly in the United States as the world's largest donor of food aid by far, are enormous. Through the various channels in which corporate actors have sought to influence food aid policy in the United States, these private actors have tried to protect their own economic interests. But as the literature on the impact of in-kind food aid has demonstrated, food aid tied to in-kind donations can have seriously detrimental effects in recipient countries. Not only is it more costly to provide such aid, but it also ultimately results in less food being delivered to the hungry, which ends up costing lives. Markets within recipient countries face disruption and distortion from in-kind provision of food aid, especially when it is monetized and sold on local markets. Displacement of commercial imports is another side effect of in-kind aid. Food preferences are also less likely to be met with heavily tied aid, and potential environmental consequences, such as GMO contamination of local crops, is a serious issue for many countries.

While industry has exerted its power through a variety of channels in an attempt to influence US food aid policy, and has thus far been successful in heading off major reform of that policy, the battle is not over. Proposals for reform of US food aid have arisen despite the powerful lobby presented by industry groups. With rising food prices, public debate over the issue, and pressure for reform, are likely to continue to intensify.

Acknowledgments

I would like to thank Doris Fuchs, Linda Swanston, and three anonymous reviewers for helpful comments on this chapter. I would also like to thank Dave Campanella and Justin Williams for research assistance, and the Social Science and Humanities Research Council of Canada for funding for this research.

Notes

1. The members of the Coalition for Food Aid included ACDI/VOCA, Adventist Development and Relief Agency International, Africare, American Red Cross, CARE, Catholic Relief Services, Counterpart International, Food for the Hungry International, International Orthodox Christian Charities, International Relief and Development, Land O'Lakes, Mercy Corps, OIC International, Project Concern International, Save the Children, and World Vision.

2. Murphy and McAfee (2005) point out that not all food aid is procyclical, however, with the WFP provision of aid tending to be countercyclical.

References

ActionAid International. 2005. *Power Hungry: Six Reasons to Regulate Global Food Corporations*. London: ActionAid International. http://www.actionaid.org/docs/power_hungry.pdf.

Ad Hoc Coalition in Support of Sustained Funding for Food Aid. 2007. Statement for the Hearing Record of the Subcommittee on Agriculture, Rural Development, FDA, and Related Agencies, Committee on Appropriations, U.S. House of Representatives. March 19.

Agricultural Food Aid Coalition. 2007. Food Aid Principles for the 2007 Farm Bill. May 10. http://www.namamillers.org/FoodAidPrincipals07.html.

American Maritime Officer. 2006. *2007 Budget Proposal Would Cut Food Aid Programs, Damage U.S. Agricultural and Shipping Industries: Cargoes Traditionally Reserved for U.S.-Flagged Vessels Would be Replaced by Grants*. http://www.amo-union.org/newspaper/morgue/3-2006/Sections/News/pref.htm.

Barrett, Christopher, and Daniel Maxwell. 2005. *Food Aid After Fifty Years: Recasting Its Role*. London: Routledge.

Bread for the World Institute. 2006. *Feeding a Hungry World: A Vision for Food Aid in the 21st Century*. http://www.bread.org/learn/global-hunger-issues/feeding-a-hungry-world-a.html.

Canadian International Development Agency (CIDA). 2005. *Canada Opens Food Aid Purchases to Developing Countries*. http://www.acdi-cida.gc.ca/CIDAWEB/acdicida.nsf/prnEn/JER-32714474-R82.

CARE USA. 2006. White Paper on Food Aid Policy. June 6. www.care.org/newsroom/articles/2005/12/food_aid_whitepaper.pdf.

Catholic Relief Services (CRS). 2007. CRS Testimony on Food Aid and the Farm Bill—House Foreign Affairs Committee, Subcommittee on Africa and Global

Health. May 24. http://www.crs.org/about_us/newsroom/speeches_and_testimony/releases.cfm?ID=40.

Clapp, Jennifer. 2004. WTO Agricultural Trade Battles and Food Aid. *Third World Quarterly* 25 (8): 1439–1452.

Clapp, Jennifer. 2005. The Political Economy of Food Aid in an Era of Agricultural Biotechnology. *Global Governance* 11 (4): 467–485.

Clapp, Jennifer. 2008. The Struggle for Global Food Aid Reform: The Politics of Policy Reform in the In-Kind vs. Cash Debate. Paper presented at the International Studies Association Annual Convention, San Francisco, March 26–29.

Cohn, Theodore. 1990. *The International Politics of Agricultural Trade: Canadian-American Relations in a Global Agricultural Context.* Vancouver: University of British Columbia Press.

Dugger, Celia. 2005. Food Aid for Africa Languishes in Congress. *International Herald Tribune*, October 13.

Dugger, Celia. 2007a. For the Hungry in Zambia, U.S. Law May Hinder Urgent Food Aid. *International Herald Tribune*, April 6.

Dugger, Celia. 2007b. U.S. Rethinks Foreign Food Aid. *International Herald Tribune*, April 22.

Finnerty, Lynne. 2004. WTO Negotiations Must Maintain Food Aid. *Farm Bureau News*, August 9.

Food and Agriculture Organization of the United Nations (FAO). 2007. *The State of Food and Agriculture 2006: Food Aid for Food Security?* Rome: FAO.

Friedmann, Harriet. 1993. The Political Economy of Food. *New Left Review* 197: 29–57.

Haider, Rizvi. 2006. Senate Resisting Food Aid Reforms That Could Save Money, Lives, Says Top U.S. Official. *OneWorld.net*, January 5.

Hedges, Stephen. 2007. U.S. Food Aid Reform Plan Has Stout Foes. *Chicago Tribune*, May 31.

Hillgren, Sonja. 2006. Food Aid Spat. *Farm Journal*, February 21.

Hopkins, Raymond. 1993. The Evolution of Food Aid: Towards a Development-First Regime. In Vernan Ruttan, ed., *Why Food Aid?*, 132–152. Baltimore: Johns Hopkins University Press.

International Grains Council (IGC). 2004. *IGC Annual Report.* London: IGC.

Levinson, Ellen. 2007. Written Testimony of Ellen S. Levinson before the Committee on Agriculture, Subcommittee on Specialty Crops, Rural Development, and Foreign Agriculture Programs, US House of Representatives, May 10. http://www.allianceforfoodaid.com/IssuesandPress/Testimony/tabid/84/Default.aspx.

Lowenberg, Samuel. 2008. Bush in Food Aid Fight with Congress. *The Politico*, February 6. http://www.politico.com/news/stories/0208/8378.html.

Monke, Jim, Geoffrey Becker, Ralph Chite, Tadlock Cowan, Charles Hanrahan, Jean Rawson, Joe Richardson, and Jeffrey Zinn. 2006. *The FY2007 Budget Request for the U.S. Department of Agriculture (USDA).* http://www.nationalaglawcenter.org/assets/crs/RS22382.pdf.

Murphy, Sophia. 2006. *Concentrated Market Power and Agricultural Trade.* Ecofair Trade Dialogue, Discussion Paper No.1. http://www.tradeobservatory.org/library.cfm?refid=89014.

Murphy, Sophia, and Kathy McAfee. 2005. *U.S. Food Aid: Time to Get it Right.* Minneapolis: Institute for Agriculture and Trade Policy.

Nichols, John. 2005. Critics Call for Changes to U.S. Food Aid Policy: Transcript from Television Program *Market to Market,* April 18. http://www.iptv.org/mtom/archivedfeature.cfm?fid=289.

Organization for Economic Cooperation and Development (OECD). 2005. *The Development Effectiveness of Food Aid: Does Tying Matter?* Paris: OECD.

Oxfam International. 2005. *Food Aid or Hidden Dumping: Separating Wheat from Chaff.* Oxfam Briefing Paper No. 71. http://www.oxfam.org/en/files/bp71_food_aid_240305.pdf.

Partnership to Cut Poverty and Hunger in Africa. 2008. The 2008 Farm Bill: Implications for Food Aid. Policy Brief No.3, July. Washington, D.C.

Pratt, Sean. 2006. US Wheat Associates Opposed to Food Aid Changes. *Western Producer*, February 9.

Thurow, Roger, and Scott Kilman. 2005. Farmers, Charities Join Forces to Block Famine-Relief Revamp. *Wall Street Journal*, October 26, A1.

Timmer, Peter. 2005. *Food Aid: Doing Well by Doing Good.* Washington, DC: Center for Global Development.

US Agency for International Development (USAID). 1995. *Food Aid and Food Security Policy Paper* (PN-ABU-219). Washington, DC: USAID.

US Committee on Agriculture, House of Representatives. 2005. *Food Aid Programs: Hearing before the Subcommittee on Specialty Crops and Foreign Agriculture Programs of the Committee on Agriculture—House of Representatives*, June 16.

US Government. 2005. *Overview of the President's Budget.* Office of Management and Budget. http://www.whitehouse.gov/omb/budget/fy2006/pdf/budget/overview.pdf.

US Government Accountability Office (GAO). 2007. *Foreign Assistance: Various Challenges Impede the Efficiency and Effectiveness of U.S. Food Aid.* http://www.gao.gov/new.items/d07560.pdf.

US House of Representatives Committee on Agriculture. 2005a. *Goodlatte Opposes USAID Budget Proposal*, February 15. http://agriculture.house.gov/press/109/pr05215.html.

US House of Representatives Committee on Agriculture. 2005b. *Subcommittee Reviews Food Aid Programs*, June 16. http://agriculture.house.gov/press/109/pr05616.html.

US Wheat Associates (USWA). 2007. *Wheat Letter—May 3, 2007*. Washington, DC: US Wheat Associates.

Uvin, Peter. 1992. Regime, Surplus and Self-Interest: The International Politics of Food Aid. *International Studies Quarterly* 36 (3): 293–312.

Wheat Export Trade Education Committee (WETEC), US Wheat Associates (USWA), and the National Association of Wheat Growers (NAWG). 2001. *Letter from the Wheat Export Trade Education Committee, U.S. Wheat Associates and the National Association of Wheat Growers to USTR Robert Zoellick Regarding Trade Priorities*. http://www.wetec.org/WheatWTOLettertoZoellick.ivnu.

Wheat Export Trade Education Committee (WETEC), US Wheat Associates (USWA), and the National Association of Wheat Growers (NAWG). 2003. *Testimony of Alan Lee on behalf of WETEC, USWA and NAWG to the Committee on Agriculture, U.S. House of Representatives*, June 18, 2003. http://www.wetec.org/leetestimony.ivnu.

Wheat Export Trade Education Committee (WETEC), US Wheat Associates (USWA), and the National Association of Wheat Growers (NAWG). 2006. *Keep the Food in Food Aid*. http://www.wheatworld.org/pdf/Food%20Aid%20Document%20rev%20%202-06%20USWA%20WETEC%20NAWG%20Final.pdf.

World Trade Organization (WTO). 2004. *Doha Work Programme: Decision Adopted by the General Council August 1, 2004*. WT/L/579. http://www.wto.org/english/tratop_e/dda_e/ddadraft_31jul04_e.pdf.

World Trade Organization (WTO). 2005. *Doha World Programme: Ministerial Declaration,* Adopted 18 December. WT/MIN (05)/DEC. http://www.wto.org/English/thewto_e/minist_e/min05_e/final_text_e.htm.

Zerbe, Noah. 2004. Feeding the Famine? American Food Aid and the GMO Debate in Southern Africa. *Food Policy* 29 (6): 593–608.

II

Corporations and Governance of Genetically Modified Organisms

6

Feeding the World? Transnational Corporations and the Promotion of Genetically Modified Food

Marc Williams

The application of biotechnology to agriculture, whether in the form of genetically modified (GM) crops or GM food, is a controversial issue and has been so from the inception of this technology. In the past decade, the application of biotechnology to agriculture has presented regulatory authorities at the national, regional, and global levels with a number of contentious decisions. The regulation of agricultural biotechnology intersects with environmental sustainability, food policy, and health concerns. Governance issues relevant to these policy domains have given rise to a contested and conflictual arena in which governments, legislators, private actors, and civic associations contend with conflicting scientific findings and grapple with ethical dilemmas. These are not discrete and segmented issue areas, but rather represent interconnected and overlapping regulatory problems, and major political and economic struggles are taking place on this terrain. Transnational corporations (TNCs) are key political actors in debates concerning the production and application of agricultural biotechnology. This chapter focuses on one aspect of the conflict over agricultural biotechnology and explores discursive strategies devised by agrifood corporations in the contest over the regulation of genetically modified food.

The chapter examines one of the ways firms have responded to the challenge of anti-GM activists, and is primarily concerned with ideational, or discursive, power. Essentially on the defensive, firms have promoted a specific normative framework designed to support claims that regulatory agencies should permit GM crops and GM food to be sold. Contestation over GM food is not limited to a narrow terrain; issues pertaining to food safety, food security, and environmental sustainability have been central to the political battles that have emerged in a number of countries and in international organizations. At the center of anti-GM activism is a normative discourse of risk (Wales and Mythen 2002). While opponents of genetically modified organisms (GMOs) frame their opposition in terms of environmental and health problems attendant on the use of GMOs (Buttel 2005), proponents have counterattacked through the construction of an alternative, positive framing of GMOs. Central to the corporate promotion of GM food is the construction of an alternative discourse that emphasizes economic efficiency, environmental sustainability, and food security. This chapter does not explore these competing arguments concerning the potential of GM food, although this contestation is central to the attempts to promote GM food. The chapter's central concern is with the corporate campaign to promote GM foods.

Corporate efforts to promote GM food are not undertaken solely through individual efforts, but also through industry associations. Peak industry bodies include the Council for Biotechnology Information, whose membership is drawn from leading biotechnology companies and trade associations. Other bodies with an international presence are the Biotechnology Industry Organization, which has an extensive list of members and includes companies from thirty-two countries; CropLife International, a global federation representing the plant science industry; and a network of regional and national associations in ninety-one countries, including CropLife America, which represents US developers, manufacturers, formulators, and distributors of plant science solutions for agriculture and pest management in the United States.

This chapter engages with this positive framing of GM food. It investigates the advocacy of the agrifood industry in support of GM food. While the chapter is specifically concerned with discursive power it begins from the recognition that all three facets of power—instrumental, structural, and discursive—are intricately connected. It thus engages with aspects of structural and instrumental power at various stages of the analysis. The argument is developed in three stages. The first part of the chapter is concerned with the political economy of agricultural biotechnology and explores the structural power of agricultural biotechnology firms. This is not a comprehensive analysis of developments in agricultural biotechnology but rather a concise introduction to the production, trade, marketing, and consumption of agribiotech products. The aim of this section is to situate the policy conflicts over GM food within a wider political-economic context since contestation over GM food is embedded within a broader framework. The second section explores the discursive power of TNCs through the discursive constructions of GM food propagated by agribusiness and independent analysts. Sources were drawn from publications from leading biotechnology firms including BASF, Bayer CropScience, Dow AgroSciences, Mendel Biology, Monsanto, Renessen and Syngenta. The section discusses the ways supporters of agricultural biotechnology have attempted to construct a normative consensus around GMOs. It examines food security and environmental sustainability as key normative frames devised by proponents of GM crops and GM food. In a realm of nonconsensual science the power to determine what counts as knowledge, which actors disseminate it, and the terms on which it is communicated has an enormous impact on public policymaking. This section is therefore explicitly concerned with the mobilization of consensus in support of GMOs and the power of discursive constructions. The third part of the chapter provides an introduction to private actors and their mobilizing strategies in the context of agricultural biotechnology. It examines interventions designed to promote the spread of GM crops and the consumption of GM food in Africa. It thus engages with aspects of instrumental, structural, and discursive power.

The Political Economy of Agricultural Biotechnology

Although modern biotechnology dates from the 1970s, it was only in the mid-1990s that the commercial application of gene technology to plants became part of agribusiness, with the production of a limited number of GM varieties of canola, corn, cotton, soybeans, and tomatoes produced in North America in 1996. The production, distribution, and consumption of GM food are shaped by both market and nonmarket factors. Traditional economic determinants such as industrial organization, financial incentives, returns to capital, efficiencies of scale, market access and entry, the costs and benefits of government regulation, and prices have all shaped the development of agricultural biotechnology. However, technological innovation, national interest, and market competitiveness are not the only factors that determine production and consumption. When it comes to the issue of regulation and the means by which consumers have exercised choice, it is important to highlight the ways cultural and ethical issues intersect with economic variables to shape the decisions of regulatory agencies on whether GM crops should be licensed and consumers enabled to purchase GM food.

The focus of this section is on the broader structural and policy frameworks within which contestations over GM food have arisen. It begins by examining three aspects of structural power in the global economy. The discussion of structural power below is indebted to Strange 1988 and the framework devised by Clapp and Fuchs in chapter 1 of this book. Corporations exercise structural power in the area of agricultural biotechnology through control over three structures of the global political economy: knowledge, finance, and production.[1] This enables them to shape the agenda and exercise control over networks and resources. This section explores major transformations in the agricultural biotechnology industry through a focus on the three structures just mentioned. It discusses technological innovation and the attendant power exercised by holders of technology. It then examines the evolution of the sector and ability of corporations to shape outcomes through a focus on patterns of investment and finance. This is followed by a brief overview of the structural power of corporations in the global agrifood production system. This overview of the political economy of agricultural biotechnology concludes by exploring some of the limits to corporate power through a discussion of consumer resistance (especially in Europe) to agricultural biotechnology.

Knowledge as Power: Technological Innovation

One of the key sources of power for corporate actors arises from the control over the production and dissemination of knowledge. The discovery of recombinant DNA (rDNA) in the 1970s has had a profound and far-reaching impact in agriculture, health sciences, and a number of other fields. Gene technology is applied to animals or plants either through the insertion of a new gene (transgenics) or alteration of the sequence of genes (genetic modification). The potential of the application of this new technology to agriculture has been hailed as truly revolutionary. Agricultural biotechnology opens up the possibility of producing plants and animals with specific characteristics relevant to the requirements and demands of the agrifood industry. Research and development (R&D) in biotechnology has expanded rapidly since the 1990s, and, at present, it shows no sign of slowing.

Historically, technology has revolutionized agricultural production. It is therefore no surprise that the latest round of technological change—the application of biotechnology—has had a dramatic impact on agricultural production. Technological innovation has been central to the development of modern agriculture and is at the heart of industrial farming practices. It accounts for the fact that agricultural productivity has kept pace with global population growth. In the standard account, the Malthusian prophecy has been denied largely because of the impact of technological change on food production (Sagoff 2001–2002; Trewavas 2002).[2] For some analysts GM crops are simply the latest example of the successful application of technology to agricultural development. Stephen Brush (2001–2002, 136–137) claims that "since the early twentieth century, the application of genetics has been an essential and highly successful component of agricultural development." This view that GMOs are merely the successor to previous technological developments has been echoed by policymakers. As Dan Glickman, a former US Agriculture Secretary, stated, "Biotechnology's been around almost since the beginning of time. It's cavemen saving seeds of a high-yielding plant. It's Gregor Mendel, the father of genetics, cross-pollinating his garden peas. It's a diabetic's insulin, and the enzymes in your yogurt. . . . Without exception, the biotech products on our shelves have proven safe" (quoted approvingly in Monsanto 2001). The claim that genetic engineering, although representing important scientific advances with potentially huge benefits for humanity, is nevertheless not a revolutionary breakthrough presenting new ethical and health and safety concerns has been disputed by numerous critics. Indeed, as Brush (2001–2002, 145) notes, "While other agricultural technologies have met with resistance, the opposition to further diffusion of GMOs is unprecedented in organization, scope, and effectiveness."

A focus on technological innovation presents an important introduction to the political economy of biotechnology, but overstates the importance of technical change in the development of agrifood systems. Technology is not simply an objective, physical given. Technological innovation arises in the context of social and economic systems, and the adoption of technological products is shaped by social and cultural mores and existing power relations. The commercialization of agricultural biotechnology thus has to be understood in the context of a global market defined by unequal power relations. Research into biotechnology initially developed in universities and in the laboratories of major firms. The symbiotic relationship between universities and the biotechnology industry has led to the development of sustained corporate-university partnerships (Krimsky, Ennis, and Weissman 1991), a major source of knowledge production and discursive power. These two institutions remain the main engines of growth. Furthermore, prior to the 1990s research and investment in agriculture was largely undertaken by public-sector authorities, with private-sector research a small component of the total research in both developed and developing countries. The new wave of scientific advances in the 1990s coupled with changes in economic policymaking shifted research from the public to the private sector (see Huffman and Evenson 1993; Tabor 1995). This movement was also facilitated by perceived commercial opportunities, while changes in financial incentives and instruments supported commercialization of the technological advances.

Financing Agricultural Biotechnology

Another factor contributing to corporate dominance of the field of biotechnology arises from the financial structure of the industry. Leading US life science companies began investing heavily in agricultural biotechnology in response to changes in US legislation on intellectual property. Liberalization of the financial sector provided funding for these developments, and the size of US markets for canola, cotton, maize, and soybeans ensured returns on investment (Wright and Pardey 2006). Agricultural biotechnology is part of a life science industry, and investment in this sector is characterized by strategies of horizontal and vertical integration (Newell 2003). Food conglomerates and pharmaceutical corporations can form part of the chain of a single integrated production process. The competitive pressures to recoup high R&D costs propel the industry to devise means to recoup outlays on investment. It has been estimated that the heavy concentration in the biotechnology industry favors the top companies at the expense of smaller firms. One survey estimates that the top ten companies account for 72 percent of the revenues earned by the biotechnology sector (ETC Group 2005, 1).

The late 1990s were characterized by consolidations, mergers, and acquisitions (Lesser 1998) as well as by the increasing internationalization of the seed industry. It has been estimated that between 1996 and 1999 mergers in the global seed industry totaled US$15 billion (Boyd 2003). Moreover, within the developing world, public-private partnerships provide much of the funding for biotechnology research but control remains in the hands of the private sector. The high R&D costs associated with the development of biotechnology produce a high barrier to entry. In other words, the development of biotechnology requires a level of scientific and technological infrastructure that small and medium-sized firms are unlikely to possess. This combination of technological and economic imperatives has seen a narrowing in focus of investments to a limited range of crops and major divestments of plant biotechnology divisions by life science firms (Schurman 2004).

Production and the Agrifood System

The dominant role of TNCs in the knowledge and financial structures is replicated by changes in global agricultural production that have enhanced corporate power. Before examining trends in the global production of GM crops it is, therefore, important to situate this sector within the broader agricultural production system. In other words, the production, distribution, and consumption of agribiotech products are located in a dynamic global agricultural system. The global agrifood system has a number of key characteristics that help to define the political economy of agricultural biotechnology. First, in the industrialized world the large-scale restructuring of agriculture in the twentieth century resulted in the acceleration of the substitution of capital for land and labor. The process of mechanization led to increased productivity, a lower number of agricultural workers, a decline in small farms, and the consequent growth in large farms. The transnational corporation now dominates restructuring in agriculture. Flexible corporate strategies have resulted in increasing rationalization, concentration, and centralization of firms. This concentration of capital in farming has been accompanied by the increased importance of processing and a resulting decline in farmers' share of food revenue (Bread for the World Institute 2003, 24). Rising input costs and falling commodity prices hurt small farmers and led to the rise of the agrifood corporation.

Second, agrifood production has become increasingly internationalized. This has affected both the quality and type of goods produced. It has been argued that profits in this sector "depended on larger restructuring of the post-war economy toward mass consumption, especially increased consumption of animal products and high value-added manufactured foods" (Friedmann 1993, 34). Third, agricultural production is characterized by global patterns of inequality. Unlike the industrialized world, in the developing world, especially in Africa, there is a heavy dependence on agriculture. It has been estimated that agricultural production's share of the gross domestic product of developing countries is 14 percent (Bread for the World Institute 2003, 20). Furthermore, agriculture remains a major source of employment in the South. For example, in 2000 it was the largest single employment sector in the developing world, accounting for 55 percent of the total workforce (Bread for the World Institute 2003, 20). Another key feature of the global agrifood system is the impact of agricultural policies. A combination of tariffs, price and income-support schemes, and domestic subsidies in advanced industrial countries protect farmers in these countries from international competition. Agricultural lobbies are politically influential in the European Union, the United States, and Japan. The production, distribution, and consumption of GM foods have developed within this context of North-South inequality, developed-country agricultural policies, and the dominance of the agrifood system by transnational corporations.

It is within this capital-intensive, internationalized, and unequal agricultural system that global production of GM crops has been developed. However, despite dominance of the knowledge and financial biotechnology sectors, and a leading role in global agricultural production, the goals of agrifood corporations in promoting the production of GM crops have not met with unparalleled success. As Falkner shows in chapter 8 of this volume, the commercialization of GM crops has not been an unambiguous success. On the one hand, the aggregate area devoted to GM crops has continued to expand with a resulting increase in revenues from planting such crops. The proponents of GMOs emphasize the rate of adoption of transgenic crops, which they claim has been astounding. It has been argued that "in ten years biotech will be the dominant paradigm in commodity agriculture" (Giddings 2006, 274). Industry specialists point out the rapid spread in acreage devoted to such crops since 1996. For example, while in 1996–2001 the total area planted worldwide was 1.7 million hectares (mha), by 2007 total acreage devoted to such crops had increased to 114.3 mha (James 2007). The rate of increased acreage is further evident if we note that there was an increase of 24.3 mha over production in 2005, which stood at 90 mha (James 2005). On the other hand, three features of the global industry present a less dynamic picture. In the first place, this aggregate increase has not been matched by significant geographic expansion. In other words, despite the significant growth in the total acreage by 2007, only twenty-three countries were cultivating transgenic crops: the United States, Argentina, Brazil, Canada, India, China, Paraguay, South Africa, Uruguay, the Philippines, Australia, Spain, Mexico, Colombia, Chile, France, Honduras, Czech Republic, Portugal, Germany, Slovakia,

Romania, and Poland[3] (James 2007). Thus, in many respects growth has not met the aims of the industry given the various campaigns to extend the geographic reach of GM crops.

A second limitation on the expansion of GM crops arises from the concentration of production, which has remained relatively stable. Commercial production is concentrated in a few countries and is dominated by the United States. In 1999, the top four countries planting GM crops were the United States (28.7 mha), Argentina (6.7 mha), Canada (4.0 mha), and China (0.3 mha) (James 2001). In 2007, the distribution was the United States (57.7 mha), Argentina (19.1 mha), Brazil (15.0 mha), and Canada (7.0 mha) (James 2007). The United States remains the leading producer, although its share of global biotech crops has declined from approximately 68 percent in 1999 to 50 percent in 2007. And a third factor that has slowed global production arises from the limited product range developed, and authorized on a commercial basis. To date, gene technology has been applied to a limited range of crops, with four crops—canola, cotton, maize, and soybeans—accounting for almost the entire global production. In 2007, maize accounted for 47 percent of the global biotech crop area, soybeans for 37 percent, cotton for 13 percent, and canola for 3 percent (James 2007). The dominant trait since 1996 has been herbicide tolerance. Insect resistance follows, and is followed in turn by stacked genes for the two properties—that is, a combination of herbicide tolerance and insect resistance. Herbicide tolerance in 2005 accounted for 71 percent (63.7 mha) (James 2005, 4–5).

The strategies of firms promoting GMOs, as well as the opposition to GMOs by environmentalists, consumer advocates, and other groups, have to be understood in the context of the changing features of the global political economy discussed above. In the first place, investment in GMOs is predicated on reaping profits, which is dependent on increasing market share in order to meet the high costs associated with GMO research and development. Second, GMOs enter an agrifood system characterized by increasing centralization and concentration, global inequalities, and distorted market structures. The next section focuses on resistance to GM foods, since this has been an important feature of the political economy of GM foods. It discusses a counter-hegemonic response to corporate power. It is this resistance to the marketing and dissemination of GM foods that creates the necessity for the development of corporate strategies designed to promote GM food.

Consumption and Resistance

Expanded production is important, with firms reaping revenues from the sale of seeds.[4] But increased global production in the absence of higher consumption places limits on capital accumulation. Increased production has not, so far, altered the resistance of many consumers to GM food. The controversial nature of GM products has been visible in protests over the planting of GM crops, consumer resistance (especially in Europe) to GM foods, and trade disputes over the application of the precautionary principle. Thus, analysis of the role of consumers is central to explanation of the economic prospects of GM food. The data available indicates that consumers in some important markets do not accept the claim of substantial equivalence, and do perceive the influence of agricultural biotechnology as a potential risk detrimental to health. Opposition to GM foods has been evident in campaigns and in purchasing behavior. This focus on consumption and resistance highlights the role of consumers as both market actors and political actors. As market actors, consumers exhibit a reluctance or unwillingness to buy GM food on the basis of price, taste, and assessment of risk. As political actors, consumers approach GM food individually and collectively (through social movement organizations) on the basis of collective strategies of mobilization. The dominant political debate centers around discourses of risk and food safety.

To understand the political economy of food, it is therefore essential to go beyond analyses of price and income factors. An analysis has to investigate other influences on consumer preferences such as values, taste, and social attitudes. The reaction of consumers to gene technology has not followed the predictions that can be derived from a neoclassical economic model. In other words, consumers are not driven solely by price signals. Furthermore, cultural and political contexts are important, as attested by, for example, the different reactions of consumers in Europe and in North America. At the risk of oversimplification, it is possible to distinguish between skeptical, anti-GM European consumers and more receptive North American consumers. European consumers have expressed greater concern about the health risks and environmental consequences of agricultural biotechnology than their North American counterparts. Thus, while producers spend billions on advertising and marketing in an attempt to influence consumer tastes, this does not always succeed.

Proponents and opponents of GM food engage in power struggles at multiple sites.[5] Resistance to GM products is located in national, regional, and international contexts. The debates at the World Trade Organization (WTO) provide an interesting example of the intersection of trade, finance and production concerns. On the one hand, proponents of agricultural biotechnology support the WTO rules-based regime since under WTO rules product rather than process differences are central. In other words, the WTO is concerned with product differentiation rather than with the processes used to make a product. From a WTO perspective, GM products can be classified as substantially equivalent to conventional products. The agricultural biotechnology industry supports this interpretation because it outlaws discrimination against GM products on health or environmental grounds based on the production process. On the other hand, opponents of GM food argue that GM products are substantially different from conventional foods and pose possible risks to consumers. These competing views have led to the development of varying national regulations reflecting different views of risk assessment, and thus potentially contributing to market closure and trade disputes. In this dispute supporters of market-based as against risk-based assessments contend that GMOs are inherently competitive, and that barriers to trade in GMOs represent old-fashioned protection. Various studies have been conducted showing that this form of protectionism is harmful to producers (Berwald, Carter, and Gruère 2006), or that the impact of trade in GM products is welfare enhancing for all states involved (Sobolevsky, Moschini, and Lapan 2005). The impact of consumer resistance and its linkage with market access, trade politics, and trade friction is highlighted through the EU-US trade dispute over GM products (Lieberman and Gray 2006; Young 2003).

The promotion of GM food, as I have argued, is situated within an evolving political economy of agricultural biotechnology. Key features of this sector include an oligopolistic structure with the dominance of a limited number of corporations, expanded production of transgenic crops but limited geographic coverage, and significant consumer resistance to GM foods. The search for regulatory approval and attempts to gain consumer confidence develop in this wider context. In the next section I discuss one of the strategies employed by agrifood corporations in their efforts to promote the consumption of GM food.

Framing Agricultural Biotechnology: Norms and GM Food

In this section I employ the concept of frame to discuss the discursive strategies of leading agricultural biotechnology firms. The concept of frames—originating in Erving Goffman's (1974) work on the ways individuals negotiate everyday life—has been given wider application in the social sciences. The concept has been used extensively by sociologists to analyze the roles of social movements, and to study collective action (Benford and Snow 2000; McAdam, McCarthy, and Zald 1996). Constructivist scholars in the field of international relations have imported it to study subjects as diverse as transnational advocacy networks (Keck and Sikkink 1998) and peace negotiations (Barnett 1999). Frames have been defined as "action-oriented sets of beliefs and meanings that inspire and legitimize the activities and campaigns of a social movement organization" (Benford and Snow 2000, 614). They can be understood as "an active, processual phenomenon that implies agency and contention" (p. 614). Framing is a useful concept applicable to the behavior of TNCs in the struggle over the acceptance of GM foods. TNCs exhibit a shared understanding of the science of agricultural biotechnology, attribute resistance to biotech products to identifiable movements and attitudes, articulate a preferred set of regulatory instruments, and act in concert with others to secure governing arrangements conducive to the promotion of GM products. The term *frame* is thus used here to refer to the scientific and normative assumptions and prescriptions embodied within a given approach to a social issue. In other words, frames define technical and social responses to problems.

Since its introduction, agricultural biotechnology has been framed in a number of different ways.[6] I will explore two central frames used by agribusiness and supporters of gene technology in agriculture: food security and environmental sustainability. These frames are embedded in three "myths" or narrative constructions devised by the agrifood industry to expand production, foster trade liberalization, gain consumer confidence, and facilitate regulatory approval. These "myths" are evident in the publications emanating from the agbiotech firms. Of course, they are portrayed as functional imperatives. Such a conceptualization thus places them as functionally necessary and as technical features rather than as competing paradigms.

Leading firms in the agrifood industry emphasize three imperatives behind agricultural biotechnology: (1) improved efficiency, (2) the eradication of world hunger, and (3) the enhancement of the nutritional intake of the developing world. The first imperative—improved efficiency—is used to construct a position in support of environmental sustainability.[7] Not only does biotechnology promise to improve plant effectiveness, but it may also produce benefits that far outweigh conventional breeding technologies. Given that innovation and science are at the cutting edge of production efficiency, firms are investing in research in agricultural biotechnology to improve their contribution to sustainable production and consumption.

The eradication of world hunger is explicitly linked to global population policy and the carrying capacity of the earth. Increasingly, agricultural biotechnology is sold as the answer to world hunger.[8] The demand for an end to world hunger is held to be partly responsible for a switch to modern seed varieties. The third imperative is health-driven, and presents agricultural biotechnology as an important contributor to global nutrition.[9] Through improvements in the nutritional components of food, GM food can assist in eradicating malnutrition or preventing blindness. These three imperatives are then used to construct a specific normative framework underpinning the application of plant and animal biotechnology.

The first and most persistent way that industry representatives and supporters frame agricultural biotechnology is in terms of food security, by explicitly invoking all three imperatives—efficiency, eradication of hunger, and nutritional enhancement. This frame locates the causes of hunger and malnutrition in low crop productivity, and conceptualizes gene technology as the solution to want and deprivation. The food security frame has had powerful resonance among some Third World governments, development nongovernmental organizations (NGOs), and some intergovernmental organizations such as the Food and Agriculture Organization (FAO). That this frame would have resonance is not surprising given the context of global hunger. Food security is, however, a highly contentious issue.

A second framing of agricultural biotechnology is in terms of environmental sustainability and is nested in the first imperative—that is, it relies on arguments predicated on a more efficient use of resources. An environmental sustainability frame may seem surprising since many opponents of GM crops also frame their opposition in these terms. Proponents of GM foods have countered with an alternative environmental perspective. As used in this chapter, food security and environmental sustainability are two complementary perspectives on the utility of agricultural biotechnology. Both frames attempt to discursively construct the GM debate in ways conducive to the interests of agribusiness TNCs. Both frames are based on a strong belief in the efficacy of science, economic efficiency, and private-interest governance.

Food Security

The frame of food security has been the major ideational underpinning of proponents of agricultural biotechnology. Defining agricultural biotechnology to encompass food security makes it an issue of central concern for governments in poor countries. Constructivist and critical scholars have demonstrated that security is not a thing or state of affairs that is universal or timeless in its meaning (see Booth 2005; Burke and McDonald 2007). It is, rather, a fluid and socially constructed phenomenon, the meaning of which changes significantly across time and space. Following from this observation, we can note that while there have been dominant understandings of food security, the term has no single, fixed meaning. Simon Maxwell and Marisol Smith (1992) identify two hundred definitions of food security, and Ian Scoones (2002) outlines seven perspectives on it. Proponents of agricultural biotechnology have developed a specific construction of food security in support of their contention that gene technology has a vital role to play in the developing world.

Since the mid-1970s, a number of official definitions of food security have been developed in the frameworks of multilateral institutions. In 1974, the World Food Conference placed the emphasis on food availability and supply at a global level, when it defined food security as "availability at all times of adequate world supplies of basic foodstuffs to sustain a steady expansion of food consumption, and to offset fluctuations in production and prices" (FAO 2003, 27). An influential World Bank report in 1986 shifted the focus to issues of access and individual well-being when it defined the term as "access by all people at all times to enough food for an active healthy life" (FAO 2003, 27). This shift moves food security from a concern with the "failure of agriculture to produce sufficient food at the national level" to a concern with the "failure of livelihoods to generate access to sufficient food at the household level" (Devereux and Maxwell 2001, i). Perhaps the most influential current definition is that given by the FAO. The FAO notes that "food security exists when all people, at all times, have physical and economic access to sufficient, safe and nutritious food to meet their dietary needs and food preferences for an active and healthy life" (FAO 2002). Revisiting these definitions, we can discern four aspects central to current concerns with food security: (1) adequacy of food supply or availability; (2) stability of supply, without fluctuations or shortages from season to season or from year to year; (3) accessibility to food or affordability; and (4) quality and safety of food. A fifth concern—the culturally appropriate nature of the food consumed—is also embedded in most current definitions. The frame of food security adopted by proponents of agricultural biotechnology essentially highlights the first and fourth components listed above. The second feature is addressed in the related frame of environmental sustainability, to be discussed below. This construction of food security is silent on distributional issues. I do not intend to suggest that the framing of agricultural biotechnology in terms of food security is a conspiracy devised by agribusiness TNCs! Nor do I intend to impugn the intellectual honesty and integrity of proponents of GMOs as a solution to food insecurity in the developing world. Rather, I am making a link between the food security frame and the interests of agribusiness TNCs. Moreover, the link between interest and normative commitment is furthered through the explicit support given to the food security frame by representatives of agribusiness TNCs and consultants to the industry.

Given the different definitions of food security it is unsurprising that the food security frame constructed in support of GM food is contested by various groups. It is beyond the scope of this chapter to explore alternative narrative frames. But before elaborating on the food security frame strategically adopted by proponents of GM food, I will briefly note the critical response. Critics of the food security frame contend that food insecurity, in the form of famine, arises from distributional issues rather than limited production (Altieri and Rosset 1999). Furthermore, they argue that existing inequalities in economic and political systems may be exacerbated by the introduction of GM crops, thus increasing rather than decreasing hunger (Altieri and Rosset 1999). Critics are also skeptical about the claims made for the nutritional enhancement capabilities of GM foods (see Bhat and Vasanthi 2002).

Agricultural biotechnology is framed in terms of its potential to eradicate world hunger. Proponents emphasize the existence of a future food gap based on projections of population increase, urbanization, and changing consumption patterns. The World Food Prospects model produced by the International Food Policy Research Institute (IFPRI)

(Pinstrup-Andersen, Pandya-Lorch, and Rosegrant 1999) has been influential in providing a basis for arguments concerning the need for more growth. As the IFPRI Report argues, "Meeting the needs of a growing and urbanizing population with rising incomes will have profound implications for the world's agricultural production and trading systems in coming decades" (p. 5). Gordon Conway (1999), former president of the Rockefeller Foundation, a leading funding agency of biotechnology-related research, has stated that "by 2020 there will be an extra 2 billion mouths to feed. . . . Biotechnology is going to be an essential partner, if yield ceilings are to be raised, if crops are to be grown without excessive reliance on pesticides, and if farmers on less favored lands are to be provided with crops that are resistant to drought and salinity, and that can make more efficient use of nitrogen and other nutrients." Similarly, the Indian biotechnologist C. S. Prakash claims that "through judicious deployment biotechnology can address environmental degradation, hunger, and poverty in the developing world by providing improved agricultural productivity and greater nutritional security" (AgBioWorld 1999). This articulation of a positive relationship between the application of biotechnology to agriculture, and the salience of GM crops to the global fight against hunger, is a powerful and sophisticated argument from the perspective of policymakers.

Apart from adding to the total stock of food available, it has been claimed that the application of agricultural biotechnology can also promote food security through nutritional enrichment. Representatives from the agricultural biotechnology industry contend that malnutrition can be effectively combated through the development of crops containing enhanced levels of vitamins and minerals. The most advanced transgenic crop in this regard is transgenic rice. So-called golden rice—that is, GM rice enhanced with vitamin A—has been promoted as a food with the potential to alleviate vitamin A deficiency. A senior executive of a biotech firm has claimed that "the levels of expression of pro-vitamin A that the inventors have achieved, are sufficient to provide the minimum level of pro-vitamin A to prevent blindness affecting 500,000 children annually, and to significantly alleviate Vitamin A deficiency affecting 124,000,000 children in 26 countries" (Five Year Freeze 2002, 20).

Environmental Sustainability

It is at first sight paradoxical, but on closer inspection not altogether surprising, that both critics and supporters of GM food frame their concerns in terms of environmental sustainability. In some respects this is a consequence of the plasticity of the term, but most importantly it is the result of the current status attached to environmental sustainability. It has become a preeminent norm in international relations. Since the United Nations Conference on Environment and Development in 1992, sustainable development has become a core objective of local, national, regional, and global governance. The promotion of environmentally sustainable policies is thus a fundamental aspiration of governing authorities. It is inconceivable that any state, international organization, or NGO would reject this normative consensus.

The frame of environmental sustainability promoted by proponents of GM food has been advanced in contradistinction to the arguments of critics of gene technology, who emphasize the environmental risks associated with the new technology. In identifying potential risks to the environment, these critics have been effective in mobilizing opinion against GM crops and food. Before examining the environmental sustainability frame of proponents of agricultural biotechnology, it is therefore necessary to present the arguments of the opponents of GM crops. The frame of environmental risk has been articulated in relation to the impact of transgenic biotechnology products on ecosystems and the dangers arising from genetically modified organisms escaping into the environment (Rissler and Mellon 1996). Critics claim that genetic pollution and imbalances in the prevailing order will result from the introduction of novel plants (Clark 2006). A number of risks have been identified, including loss of biodiversity (Losey, Rayor, and Carter 1999; Hansen Jesse, and Obrycki 2000) as well as adverse effects on soil function and ecosystem degradation. Furthermore, it is likely that the newly introduced plants will have advantages over the native species (Faure, Serieys, and Bervillé 2002), allowing them to triumph in the competition for resources. A related fear is the likelihood that transgenes will escape from agriculture with serious environmental consequences (Bergelson, Purrington, and Wichmann 1998). Among other things, critics of agricultural biotechnology claim that GM crops will become invasive. This will result in the development of more aggressive weedy types. The resulting superweeds, it is claimed, will reduce crop yields and cause serious disruptions to natural ecosystems and losses in biodiversity. This risk assessment is not shared by all scientists (Dale, Clarke, and Fontes 2002), and indeed a positive rather than a skeptical approach to the environmental impact of gene technology underpins the environmental sustainability frame as they have developed it.

The positive framing of agricultural biotechnology in relation to environmental sustainability begins with a rejection of the environmental risk frame. Advocates of agricultural biotechnology argue that the critics have overstated the potential impact of GM crops on ecosystems (FAO 2004). Since GM organisms are finely modified forms of existing ones, the fear of widespread systemic damage is misplaced. Moreover, given the existence of established procedures for their use, it is unlikely that harmful effects on the soil will remain undetected before preventive measures can be put in place. Similarly, claims concerning the development of superweeds are also dismissed. While outcrossing from domesticated GM crops to weedy and indigenous wild relatives remains a possibility, proponents claim that the frequency of such events will be extremely low. Furthermore, very few domesticated plants naturalize, and almost none are weeds in natural ecosystems. They argue that it is difficult to see how the traits that are currently being introduced into genetically modified organisms will improve their fitness in ways that allow these plants to pose a threat to the environment. Support for these views was given in an authoritative report prepared for the Australian government, which concluded that "there is no scientific reason to suspect that the nature of the hazard associated with virus recombination (i.e., the formation of a new virus) will differ for transgenic and non-transgenic plants. The transgenic plant carrying a virus-derived sequence presents an increase in risk compared to the non-transgenic plant only if the frequency with which viable recombinants are generated in the former is significantly greater" (CSIRO 2002, 8).

In addition to this rejection of the arguments of those critical of the environmental impact of gene technology, proponents construct a positive portrait of its likely beneficial contribution to environmental sustainability. Agricultural biotechnology, it is argued, will contribute to environmental sustainability through its revolutionary impact on crops. This new technology has the potential to effect a transformation of the natural limits to crop production, thus helping countries attain the goal of sustainable development. The environmental sustainability frame emphasizes the impact of transgenic technology on existing practices and the importance of novel crop varieties. Advocates claim that agricultural biotechnology promotes the existence of linkages between increased efficiency in farming practices, leading to higher yields and to environmentally sustainable development. From this perspective, biotechnology promotes environmentally sustainable production through a reduction in the application of chemicals to the soil in the form of pesticides (Huang et al. 2003) and herbicides, since the process of transgenic insertion leads to the development of crops with reduced dependency on fertilizers and water, and with greater resistance to pests, disease, and drought.

Furthermore, it is argued that farming methods will become more environmentally sustainable because of a resource saving in water use through the development of crops with reduced dependency on water, the introduction of crops that can grow in soils with high salinity, a lessening of soil erosion and gas emissions, and improvement in the productivity of marginal cropland. The widespread application of these techniques promises a revolutionary impact on agricultural production (Pringle 2003), since genetically modified crops can also produce higher yields and lower costs through a decrease in the application of fertilizers, herbicides, and pesticides. The biotech industry promotes research whose conclusions support this line of argument.[10]

The twin frames of food security and environmental sustainability are integrated into a campaign to promote GM crops and GM food. In this respect, the frames of food security and environmental sustainability are inextricably intertwined. As George Khachatourians (2001, 13) says, "Genetic engineering promises to make important contributions, making agriculture environmentally sustainable and providing food security." That is, GM crops will contribute to food security through the production of more food in an environmentally sustainable manner (Serageldin 1999).

From the perspective of the biotechnology industry, transgenic crops will promote food security through achieving higher yields and improving the use of marginal agricultural land, thus meeting the rising food demands of an increasing world population.[11] The food security and environmental sustainability discursive frames are both based on the recognition of limitations to improvements in agricultural productivity in terms of increased yields and improved quality using traditional methods of cultivation. To some extent, proponents invoke the Green Revolution of the 1960s as a model. The Green Revolution was a concerted international attempt to move beyond traditional methods through the application of science to agriculture. Plant biotechnology is therefore a necessary addition to traditional methods. The use of plant biotechnology can contribute to plant growth and development. It can speed up the breeding season, create more robust crops, and counter the ravages of nature on crops. In short, agricultural biotechnology can help to enhance crop yields and food production, thus alleviating world hunger. A leading biotech firm has claimed that

> applications of biotechnology in agriculture are in their infancy. . . . The rapid progress being made in genomics may enhance plant breeding to help secure better and more consistent yields. This would be of great benefit to those farming marginal lands worldwide. Arable land is disappearing and even if every acre is maximized using conventional agriculture, we start to come up short if we have to feed 8 billion people in 2025 and beyond. (Kirk 2000)

Promoting GM Food: Mobilizing in Support of Food Security and Environmental Sustainability

It is not enough to note the existence of the food security and environmental sustainability frames. Therefore, in this section I will briefly indicate some of the ways TNCs have utilized discursive forms of power to promote the adoption of transgenic crops and GM food in Africa. Africa is used as an example because its mass poverty, failure to gain from the Green Revolution (Paarlberg 2006), continued agricultural stagnation, and low adoption of GM crops make it a prime target for the "feeding the world" and "environmental sustainability" discourses. Only one African country, the Republic of South Africa, is currently producing GM crops. This limited penetration of African agriculture by transgenic crops is the result of a number of factors. These include the absence of incentives for private-sector investment such as poorly functioning seed markets, limited protection for intellectual property, and underdeveloped biosafety regulatory systems. In addition to these market characteristics are ethical concerns surrounding environmental risk as well as human health and safety. Thus Africa represents untapped potential for the commercialization of GM crops, and a site of struggle over the acceptance of GM crops and GM food. Large corporations have been central to the evolving history of agricultural biotechnology in Africa.

As has been noted previously, power is not a singular commodity, but is expressed in three different forms: structural, instrumental, and discursive. Thus, before exploring the mobilization of support for specific narratives of GM crops and food—that is, those supportive of the food security and environmental sustainability frames—I will briefly indicate the importance of structural and instrumental forms of power in this context.

The leading biotech firms exercise structural power through their ability to shape market developments. Important in this context is their control over the finance, production, and knowledge structures. African countries and firms lack the financial resources required to develop independent GM crops. Thus they are dependent on the dominant firms in the industry. Furthermore, the market environment provides weak incentives for private sector investment in crops of relevance to Africa, and other developing regions (Traxler 2006). Thus, R&D efforts have focused on canola, maize, and soybeans—crops of importance to temperate-zone, industrialized countries.[12] The corporate sector has ignored food crops with relevance to Africa, like cassava (Fukuda-Parr 2006). One example of the agenda-setting power of the corporate sector has been in the debate over food aid. It has been argued that agribusiness TNCs have formed alliances with governmental agencies to foster the export of GM crops. For instance, in mid-2002 the United States sent 500,000 tons of corn as food aid to Southern Africa (Lesotho, Mozambique, Swaziland, Zambia, and Zimbabwe) in response to fears that the region was on the verge of a famine. Seventy-five percent of the shipment contained GMOs, although non-GM grain was available in sufficient quantities (Greenpeace 2002).

The activities of leading biotech firms affect the food security and sustainable development agendas of developing countries through a variety of mechanisms. Osgood (2006) notes five modalities through which firms exercise instrumental and discursive forms of power. These are philanthropy, market development, partnerships with public-sector and civil society groups, knowledge sharing, and advancing regulatory structures and biosafety regimes. Corporate philanthropy donates activities such as gifts of cash, technical training, and research materials. Closely related to philanthropy is knowledge sharing, which incorporates the dissemination of scientific knowledge—in other words, making research results available to the wider scientific community. Market development refers to the development of joint-venture commercial strategies for small-scale farmers. Partnership with public or civil society groups to sponsor joint research activities and crop trials with governments and research institutes in the developing world utilizes a firm's control over knowledge to foster commercial interests. Such behavior is at the intersection of instrumental and discursive forms of power. One example of such collaboration occurred in Kenya. In September 2001, in conjunction with the Kenya Agricultural Research Institute (KARI), the University of Missouri, and the US Agency for International Development (USAID), Monsanto began testing a sweet potato variety engineered to have increased resistance to insects (Mackey 2002). A representative of the Kenyan agricultural ministry, referring to the debate on the acceptability of GM crops, claimed that "biotechnology is not our problem. Poverty is" (Mugai 2000). The seeds were initially donated by Monsanto, with the aim of aiding food security without jeopardizing future business opportunities, particularly in the developed world (Wambugu 2001).

The Kenya sweet potato project is one example of claims made by Monsanto representatives that genetically modified foods improve human nutrition, both by increasing the nutritional content of foods and by enhancing food security (Mackey 2002, 157). Claims that biotechnology is necessary to ensure food security in Africa are also made by the director of the Africa regional office of the International Service for the Acquisition of Agro-Biotech Applications (ISAAA), Florence Wambugu (see Wambugu 1999, 15–16). The claims about the sweet potato's ability to ease the problems of famine in Kenya have been refuted in a number of places (Beingessner 2003; Greenpeace Genetix Crime Unit 2008). But in this and other cases, firms have been active in presenting their views to governments about the desirability and enforceability of different approaches to the regulation of GMOs. The economic and strategic potential of biotechnology places firms in a strong position to assert their preferences regarding the scope of regulations, the speed of the process, and the nature of the risks they address. This material power translates into high levels of interaction with governments through active consultations and membership on committees.

Furthermore, firms have also exercised discursive power through their ability to promote their work as the key to broader governmental objectives such as promoting growth and poverty alleviation. Some of the controversial claims made by companies about the merits and pro-poor nature of the technology are being increasingly internalized by academics and policymakers. To illustrate the discursive power of the food security and environmentally sustainable frames I will examine the emerging consensus on food security, sustainable development, and agricultural biotechnology in Africa.

Three recent publications suggest the emergence of a consensus on the benefits of agricultural biotechnology for Africa. Norman Clark, John Mugabe, and James Smith (2005) situate the need for agricultural biotechnology firmly in the context of food insecurity. They claim that "persistent poor agricultural production and rising food insecurity in Sub-Saharan Africa have brought into sharp focus the role of modern agricultural biotechnology in human development" (p. 1). The authors accept an important role for agricultural biotechnology as part of the solution to Africa's food insecurity.[13] The issue is not whether biotechnology is the solution but rather how African countries can best benefit from these technological advances. Thus the authors conclude that "for Africa to benefit from the rapid scientific and technological advances associated with modern agricultural biotechnology its countries need to build public confidence in the role of technology and its implications for human development" (p. 33). Similarly, in the draft report prepared by a High-Level Panel of Experts convened by the African Union (AU) and the New Partnership for Africa's Development (NEPAD), the key problems are presented as "long-term issues of hunger, nutrient deficiency, and threats to overall agricultural productivity caused by unfavorable climate, diseases and soil infertility" (AU/NEPAD 2006, 2). To this end, African countries are urged to "invest in agricultural biotechnology projects and capacity-building" (p. 2) to address these serious issues. A third recent study to focus on these issues reaches similar conclusions. The authors examine the poor state of African agriculture, a result, they argue, of natural conditions and poor policy choices. They conclude that "agricultural biotechnology offers the potential to increase yields, enable adaptation to more extreme environments, and improve nutritional content" (Southgate and Graham 2006, 1). They argue that this outcome is unlikely to be realized if the claims of the anti-GM activists are not countered.

While debate on the contribution of transgenic crops to environmental sustainability is far from conclusive, recent evidence suggests greater attention is being given to the environmental sustainability frame. On the one hand, a study of the potential for GM crops to contribute to poverty alleviation in Africa concluded that the environmentally sustainable consequences of transgenic cotton, maize, and sweet potato is low to moderate (de Grassi 2003). The AU/NEPAD study, on the other hand, notes that "new technology, especially biotechnology, green chemistry and nanotechnology will drive greater eco-efficiency, resource productivity and a paradigm shift across the economy" (AU/NEPAD 2006, 28). Such optimism is echoed by Paarlberg (2006, 91), who notes the potential for drought-resistant strains of maize, sorghum, wheat, rice, or millet to offer "poor African farmers something far more valuable than the insect resistance or herbicide resistance traits of the first generation GM crops."

Conclusion

This chapter has examined the activities of private actors in the debate over GM food. It has focused on the construction of ideas about biotechnology, which has been an important part of the contestation between GM and anti-GM campaigners. The normative frames of food security and environmental sustainability are strategies devised by private actors to influence the regulatory environment. Agricultural biotechnology firms have been active at the national, regional, and global levels in attempting to shape the governance of agricultural biotechnology.

The first section of the chapter discussed the ways structural and instrumental power is exercised in the global political economy and the role of biotech firms in this power structure. The second section explored the discursive power of agricultural biotechnology companies in promoting GM food. In discussing the role of discursive power in constructing ideas about Africa and biotechnology, the third section linked the exercise of discursive power with the structural and instrumental forms of power available to corporate actors.

By promoting GM crops in terms of their potential to contribute to the important ethical and political goals of ending world hunger, TNCs have attempted to reframe agricultural biotechnology as a food security rather than a food safety issue. This framing has less immediate relevance for populations in advanced industrial countries, where (on the whole) the central issues for farmers and consumers is glut rather than hunger, but it is vital for parts of the developing world in which food availability is a salient topic. TNCs have therefore attempted to mobilize support for GM crops among governments, scientists, farmers, and consumers in developing countries.

Moreover, in terms of the promotion of sustainable production in contradistinction to environmental risk, the frame of environmental sustainability resonates in both the developed and developing worlds. On the one hand, the environmental sustainability frame developed in support of agricultural biotechnology runs counter to the claims of anti-GM protesters. On the other hand, the rhetorical commitment to sustainable development accepted by the international community at the Rio

Conference in 1992 has shaped the response of private actors to issues around economic growth and development.

The production, trade, marketing, and consumption of GM products present regulatory agencies with economic, ethical, and political problems. The agribiotech industry has been mired in controversy from the outset, with contending theories and competing empirical evidence forming an inescapable context for the development of this industry. The commercialization of the technical innovations devised in laboratories is dependent on a receptive economic environment. In other words, investors will not invest unless they are assured of adequate returns on their capital outlay. One of the key strategies utilized by TNCs in the promotion of GM food is a specific normative framework conducive to the spread of agricultural biotechnology. Firms have been forced to engage on this terrain because GMOs remain deeply controversial. The production and consumption of GM crops depend on enabling legislation at the national and international levels. The food security and environmental sustainability frames are designed to create a favorable policy environment.

Acknowledgments

I would like to thank Zsofi Korosy for research assistance.

Notes

1. Strange (1988) identified four primary structures of the global political economy: knowledge, production, finance, and security. The security structure is of limited relevance to this study.

2. However, more critical accounts contend that the blind pursuit of agricultural productivity through technological innovation produces environmental harm. Recent research has demonstrated the contribution of environmentally sensitive, low-cost technologies to improved food production (see, for example, Pretty, Morison, and Hines 2003).

3. Listed in declining order of importance as producers.

4. See the analysis by Sell (chapter 7, this volume) for further analysis of corporate power and the political economy of seeds.

5. See Smythe (chapter 4, this volume) for a detailed exploration of food standards and the Codex Alimentarius.

6. My concern is with the strategies adopted by agribusiness, and not with all frames that have been advanced in the contest over GMOs. See Sell (chapter 7, this volume) for a discussion of alternative farming devices.

7. According to Dow AgroSciences (2007), "Plant biotechnology . . . provides a tool for producing food in a more sustainable manner."

8. CropLife (2006) notes that "providing sufficient, healthy food for (this) increasing population is one of the main challenges facing sustainable agriculture today. By balancing environmental, economic and social concerns, the plant science industry promotes sustainable agriculture, ensuring stable and secure food production and contributing to poverty alleviation."

9. Monsanto (2007) claims the company is "leading the way to develop agricultural products for healthier foods."

10. "Biotech is allowing farmers to practice more conservation tillage" (Linda Thrane, executive director of the Council for Biotechnology Information; quoted in Ingwersen 2002).

11. "One of the reasons we did the study was for the ag and the non-ag audience to recognize that there are some benefits from biotechnology that are not readily apparent. Improved water quality, less treatment costs at the water treatment plant, more wildlife when they're driving in the countryside—those are some of the intangibles, if you will" (Dan Towery of the Technology Information Center; quoted in Ingwersen 2002).

12. Given the diversity of the developing world, it is important to note significant exceptions. As detailed above, Argentina is the world's second largest cultivator of transgenic crops. See Newell (chapter 9, this volume) for an extended analysis of the governance of GMOs in Argentina.

13. This position is challenged by alternative perspectives claiming that similar gains in productivity can be achieved at lower environmental and social costs.

References

AgBioWorld. 1999. Scientists in Support of Agricultural Biotechnology. Text of a Petition. www.AgBioWorld.org/petition.html.

Altieri, Miguel A., and Peter Rosset. 1999. Ten Reasons Why Biotechnology Will Not Ensure Food Security, Protect the Environment and Reduce Poverty in the Developing World. *AgBioForum* 3 (3/4): 155–162.

AU/NEPAD. 2006. *Freedom to Innovate: Biotechnology in Africa's Development.* Draft Report of the High-Level African Panel on Modern Biotechnology of the African Union (AU) and the New Partnership for Africa's Development (NEPAD). http://www.nepadst.org/doclibrary/pdfs/abp_july2006.pdf.

Barnett, Michael. 1999. Culture, Strategy and Foreign Policy Change: Israel's Road to Oslo. *European Journal of International Relations* 5 (1): 5–36.

Beingessner, Paul. 2003. Sweet Potato Lie: Agents of the New Colonialism Are Using Fabrications (in More Ways Than One) to Impose GM Crops on Farmers and Consumers—. *Briarpatch Magazine* 32 (9): 16.

Benford, Robert D., and David A. Snow. 2000. Framing Processes and Social Movements: An Overview and Assessment. *Annual Review of Sociology* 26 (1): 611–639.

Bergelson, Joy, Colin B. Purrington, and Gale Wichmann. 1998. Promiscuity in Transgenic Plants. *Nature* 395 (6697): 25.

Berwald, Derek, Colin A. Carter, and Guillaume P. Gruère. 2006. Rejecting New Technology: The Case of Genetically Modified Wheat. *American Journal of Agricultural Economics* 88 (2): 432–447.

Bhat, Ramesh V., and S. Vasanthi. 2002. Can Golden Rice Eradicate Vitamin A Deficiency? *The Hindu*, December 5.

Booth, Ken, ed. 2005. *Critical Security Studies and World Politics*. Boulder: Lynne Rienner.

Boyd, William. 2003. Wonderful Potencies? Deep Structure and the Problem of Monopoly. In Rachel A. Schurman and Dennis D. K. Kelso, eds., *Engineering Trouble: Biotechnology and Its Discontents,* 24–62. Berkeley: University of California Press.

Bread for the World Institute. 2003. *Agriculture in the Global Economy.* Washington, DC: Bread for the World Institute.

Brush, Stephen B. 2001–2002. Genetically Modified Organisms in Peasant Farming: Social Impact and Equity. *Indiana Journal of Global Legal Studies* 9 (1): 135–162.

Burke, Anthony, and Matt McDonald, eds. 2007. *Critical Security in the Asia-Pacific*. Manchester: Manchester University Press.

Buttel, Frederick H. 2005. The Environmental and Post-Environmental Politics of Genetically Modified Crops and Foods. *Environmental Politics* 14 (3): 309–323.

Clark, E. Ann. 2006. Environmental Risks of Genetic Engineering. *Euphytica* 148 (1–2): 47–60.

Clark, Norman, John Mugabe, and James Smith. 2005. *Governing Agricultural Biotechnology in Africa: Building Capacity for Policy-Making* (Centre for Technology Studies). www.nepadst.org/doclibrary/pdfs/agribiotech2006.pdf.

Conway, Gordon M. 1999. GM Foods Can Benefit the Developing Countries. http://www.monsanto.co.uk/news/ukshowlib.phtml?uid=1667.

CropLife America. 2006. CropLife America Profile Brochure. http://www.croplifeamerica.org/design_06/viewer.asp?pageid=98&keyword=.

CSIRO. 2002. *Environmental Risks Associated with Viral Recombination in Virus Resistant Transgenic Plants. Final Report*. Canberra: CSIRO.

Dale, Phillip J., Belinda Clarke, and Eliana M. Fontes. 2002. Potential for the Environmental Impact of Transgenic Crops. *Nature Biotechnology* 20 (6): 567–574.

de Grassi, Aaron. 2003. *Genetically Modified Crops and Sustainable Poverty Alleviation in Sub-Saharan Africa: An Assessment of Current Evidence.* Third World Network–Africa. http://allafrica.com/sustainable/resources/view/00010161.pdf.

Devereux, Stephen, and Simon Maxwell, eds. 2001. *Food Security in Africa.* London: IT Publications.

Dow AgroSciences. 2007. *Why Biotech Is Important.* Plant Genetics & Biotechnology. http://www.dowagro.com/pgb/intro/why/.

ETC Group. 2005. *Oligopoly Inc: Concentration in Corporate Power 2005.* Communiqué Issue No. 91, November/December. http://www.etcgroup.org/upload/publication/pdf_file/44.

FAO. 2002. *The State of Food Insecurity in the World 2001.* Rome: FAO.

FAO. 2003. *Trade Reforms and Food Security: Conceptualizing the Linkages.* Rome: FAO.

FAO. 2004. *The State of Food and Agriculture 2003–2004. Agricultural Biotechnology: Meeting the Needs of the Poor?* Rome: FAO.

Faure, Nathalie, Hervé Serieys, and André Bervillé. 2002. Potential Gene Flow from Cultivated Sunflower to Volunteer, Wild Helianthus Species in Europe. *Agriculture, Ecosystems and Environment* 89 (3): 183–190.

Five Year Freeze. 2002. *Feeding or Fooling the World: Can GM Really Feed the Hungry?* London: Five Year Freeze. http://www.gmfreeze.org/pdf/Feed_Fool_World.pdf.

Friedmann, Harriet. 1993. The Political Economy of Food: A Global Crisis. *New Left Review* 197: 29–56.

Fukuda-Parr, Sakiko. 2006. Introduction: Global Actors, Markets and Rules Driving the Diffusion of Genetically Modified (GM) Crops in Developing Countries. *International Journal of Technology and Globalisation* 2 (1/2): 1–11.

Giddings, L. Val. 2006. Whither Agbiotechnology? *Nature Biotechnology* 24 (3): 274–276.

Goffman, Ervin. 1974. *Frame Analysis: An Essay on the Organization of the Experience.* New York: Harper Colophon.

Greenpeace. 2002. USAID and GM Food Aid. Greenpeace UK. http://www.greenpeace.org.uk/MultimediaFiles/Live/FullReport/5243.pdf.

Greenpeace Genetix Crime Unit. 2008. *Monsanto's 7 Deadly Sins.* Greenpeace/Hardouin. http://www.greenpeace.org/raw/content/international/press/reports/7-deadly-sins.pdf.

Hansen Jesse, Laura C, and John J. Obrycki. 2000. Field Deposition of Bt Transgenic Corn Pollen: Lethal Effects on the Monarch Butterfly. *Oecologia* 125 (2): 241–248.

Huang, Jikuang, Ruifa Hu, Carl Pray, Fangbin Qiao, and Scott Rozelle. 2003. Biotechnology as an Alternative to Chemical Pesticides: Study of Bt Cotton in China. *Agricultural Economics* 29 (1): 55–67.

Huffman, Wallace, and Robert E. Evenson. 1993. *Science for Agriculture: A Long-Term Perspective*. Ames: Iowa University Press.

Ingwersen, Julie. 2002. Biotech Soybeans Help Soil Quality, Industry Says. *Reuters*, February 22. http://www.planet.ark.org/dailynewstory.

James, Clive. 2001. *Global Review of Commercialized Transgenic Crops: 2001*. ISAAA Briefs No. 24: Preview. http://www/isaaa.org.

James, Clive. 2005. *Global Status of Commercialized Biotech/GM Crops: 2005*. ISAAA Briefs No. 34: Executive Summary. http://www.isaaa.org.

James, Clive. 2007. *Global Status of Commercialized Biotech/GM Crops: 2007*. ISAAA Briefs No. 37: Executive Summary. http://www.isaaa.org.

Keck, Margaret, and Kathryn Sikkink. 1998. *Activists beyond Borders: Advocacy Networks in International Politics*. Ithaca: Cornell University Press.

Khachatourians, George G. 2001. How Well Understood Is the "Science" of Food Safety? In Peter W. B. Phillips and Robert Wolfe, eds., *Governing Food: Science, Safety and Trade*,14–23. Montreal and Kingston: McGill-Queen's University Press.

Kirk, F. William. 2000. The 21st Century—An Agribusiness Odyssey. D. W. Brooks Lecture, University of Georgia, October 2. http://www.caes.uga.edu/events/dwbrooks/2000/kirklecture.html.

Krimsky, Sheldon, James G. Ennis, and Robert Weissman. 1991. Academic-Corporate Ties in Biotechnology: A Quantitative Study. *Science, Technology & Human Values* 16 (3): 275–287.

Lesser, William. 1998. Intellectual Property Rights and Concentration in Agricultural Biotechnology. *AgBioForum* 1 (2): 56–61.

Liberman, Sarah, and Tim Gray. 2006. The So-Called "Moratorium" on the Licensing of New Genetically Modified (GM) Products by the European Union: 1998–2004: A Study in Ambiguity. *Environmental Politics* 15 (4): 592–609.

Losey, John E., Linda S. Rayor, and Maureen E. Carter. 1999. Transgenic Pollen Harms Monarch Larvae. *Nature* 399 (6733): 214.

Mackey, Maureen. 2002. The Application of Biotechnology to Nutrition: An Overview. *Journal of the American College of Nutrition* 21 (90003): 157S–158S. http://www.jacn.org/cgi/reprint/21/suppl_3/157S.

Maxwell, Simon, and Marisol Smith. 1992. Household Food Security: A Conceptual Review. In Simon Maxwell and Timothy Frankenburger, eds., *Household Food Security: Concepts, Indicators, Measurements: A Technical Review*, 1–72. New York and Rome: UNICEF and IFAD.

McAdam, Doug, John D. McCarthy, and Mayer N. Zald, eds. 1996. *Comparative Perspectives on Social Movements: Opportunities, Mobilizing Structures and Framing*. Cambridge: Cambridge University Press.

Monsanto. 2007. Corporate Social Responsibility: Healthier Foods. Monsanto. http://www.monsanto.com/responsibility/our_pledge/facing_challenges/healthy_foods.asp.

Monsanto Biotech Knowledge Center. 2001. *Biotech Basics.* http://www.biotechknowledge.com/biotech/bbasics.nsf.

Mugai, Naftali. 2000. Transgenic Sweet Potato Could End Kenya Famine. *Environment News Service*, September 15. http://forests.org/archive/africa/trswpotc.htm.

Newell, Peter. 2003. Globalization and the Governance of Biotechnology. *Global Environmental Politics* 3 (2): 58–59.

Osgood, Diane. 2006. Living the Promise? The Role of the Private Sector in Enabling Small-Scale Farmers to Benefit from Agro-Biotech. *International Journal of Technology and Globalisation* 2 (1/2): 30–45.

Paarlberg, Robert. 2006. Are Genetically Modified (GM) Crops a Commercial Risk for Africa? *International Journal of Technology and Globalisation* 2 (1/2): 81–92.

Pinstrup-Andersen, Per, Rajul Pandya-Lorch, and Mark W. Rosegrant. 1999. *World Food Prospects: Critical Issues for the Early Twenty-First Century: IFPRI Policy Report.* Washington, DC: IFPRI.

Pretty, J. N., J. I. L. Morison, and R. E. Hines. 2003. Reducing Poverty by Increasing Agricultural Sustainability in Developing Countries. *Agriculture, Ecosystems & Environment* 98 (1): 217–234.

Pringle, Peter. 2003. *Food Inc.: Mendel to Monsanto—the Promises and Perils of the Biotech Harvest.* New York: Simon and Schuster.

Rissler, Jane, and Margaret Mellon. 1996. *The Ecological Risks of Engineered Crops.* Cambridge, MA: MIT Press.

Sagoff, Mark. 2001–2002. Biotechnology and Agriculture: The Common Wisdom and Its Critics. *Indiana Journal of Global Legal Studies* 9 (1): 13–34.

Schurman, Rachel A. 2004. Fighting "Frankenfoods": Industry Opportunity Structures and the Efficacy of the Anti-Biotech Movement in Western Europe. *Social Problems* 51 (2): 243–268.

Scoones, Ian. 2002. *Agricultural Biotechnology and Food Security: Exploring the Debate.* IDS Working Paper 145. Brighton: IDS.

Serageldin, Ismail. 1999. From Green Revolution to Gene Revolution. *Economic Perspectives: An Electronic Journal of the U.S. Department of State* 4 (2): 17–19.

Sobolevsky, Andrei, GianCarlo Moschini, and Harvey Lapan. 2005. Genetically Modified Crops and Product Differentiation: Trade and Welfare Effects in the Soybean Complex. *American Journal of Agricultural Economics* 87 (3): 621–644.

Southgate, Douglas, and Douglas Graham. 2006. *Growing Green: The Challenge of Sustainable Agricultural Development in Sub-Saharan Africa.* London: International Policy Press for the Sustainable Development Network.

Strange, Susan. 1988. *States and Markets.* London: Pinter.

Tabor, Steven R, ed. 1995. *Agricultural Research in an Era of Adjustment: Policies, Institutions, and Progress*. Washington, DC: World Bank.

Traxler, Greg. 2006. The GMO Experience in North and South America. *International Journal of Technology and Globalisation* 2 (1/2): 46–64.

Trewavas, Anthony. 2002. Malthus Foiled Again and Again. *Nature* 418 (6898): 668–670.

Wales, Corinne, and Gabe Mythen. 2002. Risk Discourses: The Politics of GM Foods. *Environmental Politics* 11 (2): 121–144.

Wambugu, Florence. 1999. Why Africa Needs Agricultural Biotech. *Nature* 400 (6739): 15–16.

Wambugu, Florence. 2001. Control of African Sweet Potato Virus Diseases through Biotechnology and Technology Transfer. *Biotechnology Seminar Paper*, ISNAR Biotechnology Service. http://www.isnar.cgiar.org/ibs/papers/wambugu.pdf.

Wright, Brian D., and Philip G. Pardey. 2006. Changing Intellectual Property Regimes: Implications for Developing Country Agriculture. *International Journal of Technology and Globalisation* 2 (1/2): 93–114.

Young, Alasdair R. 2003. Political Transfer and "Trading Up"? Transatlantic Trade in Genetically Modified Food and U.S. Politics. *World Politics* 55 (4): 457–484.

7

Corporations, Seeds, and Intellectual Property Rights Governance

Susan K. Sell

Seeds are at the center of a complex political dynamic between stakeholders. Access to seeds concerns the balance between private rights and public obligations, private ownership and the public domain, and commercial versus humanitarian objectives. Like a three-dimensional chessboard, the institutional and regulatory landscape of these politics is multilayered and multifaceted. For example, the intersection between agriculture, trade, and intellectual property governance is complicated by the diverse institutions involved, including the World Trade Organization (WTO), the World Intellectual Property Organization (WIPO), the Convention on Biodiversity (CBD), and the Food and Agriculture Association (FAO). Corporations, developing countries, and nongovernmental organizations (NGOs) deploy instrumental, structural, and discursive power, and engage in strategic forum shifting. This dynamic reveals both stark power asymmetries and opportunities. On balance, the corporations have the upper hand in this complicated game.

After 1973, a major breakthrough in biotechnology created new business opportunities. The 1973 development of the recombinant DNA technique, which enabled foreign genes to be inserted into microorganisms, launched the era of commercial biotechnology (Dutfield 2003a, 138). In 1980, the US Supreme Court ruled in *Diamond v. Chakrabaty* (447 U.S. 1980) that a human-made oil-eating bacterium could be patented. This case led to the expansion of rights to own living organisms and injected greater certainty into the development of commercial biotechnology. The ability to acquire intellectual property rights on altered life forms helped biotechnology start-up companies to raise venture capital.

Global biotechnology firms have played a prominent role in the effort to raise regulatory standards for intellectual property protection worldwide. The complexity of the regulatory environment has given powerful players important advantages in achieving their objectives. Vertical and horizontal forum shifting between multilateral, regional, and bilateral venues and across international organizations has made it far more difficult for developing countries to keep abreast of intellectual property policymaking, and has made them more vulnerable to power plays from global corporations and their supportive trade ministries such as the Office of the United States Trade Representative (USTR). This chapter examines the impact of this process on developing-country agriculture, and highlights the instrumental, structural, and discursive dimensions of the contemporary policymaking environment.

Agrifood corporations have deployed instrumental, structural, and discursive power to shift forums and join with states to influence rules—in this case shifting discussion of intellectual property into trade institutions (USTR, the General Agreement on Tariffs and Trade (GATT), and the World Trade Organization). Their *instrumental* power consists of their access to important decision-making bodies and influence over public-sector actors. Another element of their instrumental power is their ability to withhold the fruits of their invention. Even when they obtain patent rights over plant varieties, or genetic engineering tools or processes, they may choose not to license the technology. Corporations also deploy *structural* power, a more indirect form of influence. Structural power derives from their position in the seed industry. Global biotechnology firms enjoy broad property rights and economic concentration in the sector. This increases their profitability and their political power. The choices that these firms make have a significant impact on access to seeds. The third type of power is *discursive* power. Discursive power refers to the potency of the frames that actors use to couch their preferences. Both transnational corporations (TNCs) and NGOs deploy discourse as a strategy to gain power and influence policy. The discourse is highly contested and sometimes sophistry prevails over evidence-based analysis (Drahos and Tansey 2008, 203). Rivals often have equally compelling frames; in such cases it is these other dimensions of power that tend to be decisive.

NGOs and developing-country governments have also deployed structural and discursive power to shift to other forums such as the Convention on Biological Diversity (CBD) and the FAO in an attempt to achieve more favorable outcomes. While these other forums have less power as norm setters in the global economy, the contest is not over. As I will discuss in the conclusion, dynamic and promising alternatives are emerging for developing-country agriculture.

Structural Power: Private Sector Ascendant

With regard to access to seeds, the biggest change over the past century has been the shift from public provision to private provision of seeds. In the early nineteenth century, the U.S. Patent and Trademark Office (PTO) collected, cataloged, and distributed seeds freely (Aoki 2003, 331). In the United States, the government established public land grant universities to develop and provide free seeds to farmers. Farmers traditionally have saved seeds and reused them. They have traded seeds with each other, sold seeds to each other, and created and experimented with new hybrids. In these ways they have contributed to the planet's biodiversity. In the past, US laws covering plant varieties incorporated the concept of farmers' rights, in which farmers retained their freedom to engage in these important and traditional activities.

Historically, seed companies preferred to develop hybrids—that is, plants that do not breed true to type—because farmers must purchase new hybrid seeds every planting season (UNCTAD-ICTSD 2003, 107). For plant varieties lacking this built-in biological protection, plant breeders can appeal to Plant Breeders' Rights (PBRs).[1] PBRs grant a legal monopoly and give the rights holder control over "the sale, reproduction, import, and export of new varieties of plants" (Binenbaum et al. 2003, 312). PBRs have led to increased private investment in agricultural research. However, PBRs "generally do not encourage breeding related to minor crops in small markets" (UNCTAD-ICTSD 2003, 106). As a result, the private sector underinvests in crops and technologies suitable for smallholder farmers and thus these goods are underprovided.

The seed sector is characterized by marked economic concentration that has only increased in recent decades, owing to a combination of expanded intellectual property rights and relaxed antitrust enforcement (Sell 1998). In agribiotechnology, six companies alone hold 75 percent of all US patents granted to the top thirty patent-holding firms: Monsanto, DuPont, Syngenta, Dow, Aventis, and Grupo Pulsar (Dutfield 2003a, 154; Fowler 1994, 146). The top ten seed companies already control 57 percent of the global seed market (ETC Group 2008b, 1).

More recently, agrifood TNCs have presented genetically modified (GM) seeds as a solution to the problems posed by climate change.

Monsanto of the United States and BASF of Germany have entered into a US$1.5 billion partnership to engineer stress-tolerant plants (ETC Group 2008a). Agrifood TNCs have applied for over 500 patents, with patent claims extending not only to particular stress-tolerant traits in a single engineered plant species, "but also to substantially similar genetic sequence[s] in virtually all transformed plants" (ETC Group 2008b, 6). ETC notes with alarm the broad scope of these patents and the likelihood that they would extend to all flowering plants, from which virtually all of the world's food supply comes (p. 6). If patent offices grant all the gene patents that BASF of Germany, Syngenta of Switzerland, and Monsanto of the United States have applied for, these three firms will control nearly two-thirds of the climate-related gene families (Weiss 2008, A04). According to Safrin, "Some resources benefit from being shared. . . . The more the resources are shared, the more they are preserved. Genetic resources are this type of good" (2004; quoted in Rajotte 2008, 160). To the extent that this TNC "gene grab" will prevent resource sharing and lock out potential user-innovators, this will have negative effects on biodiversity, competition, and food security. Farmers will be prevented from "breeding, saving and reusing seeds to feed themselves and their communities" (Rajotte 2008, 162).

Strategic partnerships and extensive cross-licensing are key factors driving consolidation of the sector. These factors have increased both regulatory and intellectual property transaction costs. According to Krattiger (2004, 14), "Paradoxically, purchasing an entire company is nowadays often much cheaper than obtaining a license or resolving patent disputes." This combination of economic concentration with extensive and broad patenting means that a handful of global corporations are increasingly controlling the world's food supply and entangling farmers in a dense web of licensing and royalty obligations.

In the past, the public sector dominated agricultural research; today, in industrialized countries the private sector accounts for the vast majority of agricultural research. Michael Pragnell (2005, 93), former chief executive officer (CEO) of Syngenta AG, which is the largest investor in agricultural research (US$800 million per year), has stated that "such research is inevitably targeted toward the major existing markets." The current system skews research toward rich and middle-income countries' markets and sectors (Barton 2003; Lettington 2003; Rai and Eisenberg 2003). In the agriculture sector this means concentrating research and development (R&D) on crops such as cotton, soybeans, maize, and canola, and not pursuing research that would be beneficial to less lucrative microclimates (Lettington 2003). Indeed, given the large investments in agricultural R&D (e.g., Monsanto spends over US$500 million per year), to recoup their costs firms must develop crops that will be planted on a large commercial scale (Wright and Pardey 2006b, 22; Tansey 2006, 6). The increasing domination of the private sector in agriculture means that, for developing countries, significant public-sector research will be imperative to meet their needs.

Since public and nonprofit sectors will be central to agricultural R&D, crop development, and commercialization of staple foods in much of the developing world, a key question is whether these sectors will be able to gain access to the necessary technologies held by the leading global firms (Wright and Pardey 2006b, 23).

Instrumental Dimensions of Corporate Power

The instrumental power of global firms is reflected in the membership of key policymaking committees in US trade institutions. These committees assist US trade negotiators in designing policies for multilateral, regional, and bilateral trade. The USTR Agricultural Trade Advisory Committee represents these corporations: Altria, Archer Daniels Midland, Blue Diamond Growers, Burger King, Campbell Soup, Cargill, ConAgra, CropLife America, General Mills, Hershey Foods, HJ Heinz, Land O' Lakes, Louis Dreyfus, McDonald's, Monsanto, Sunkist Growers, Sun Maid Growers, Wal-Mart, and Yum! Restaurants. The USTR's Industry Trade Advisory Committee on Intellectual Property Rights includes representatives for Biotechnology Industry Organization, Eli Lilly and Company, Intellectual Property Owners, Merck & Company, Inc., Pfizer, and the Pharmaceutical Research and Manufacturers Association (PhRMA). None of these firms or organizations promotes the interests of smallholder farmers. The reach of these firms can be quite broad. For instance, the most influential NGO in Brazilian agricultural trade policy is the Institute for the Study of Trade and International Negotiations (ICONE). ICONE is supported by the Brazilian Agribusiness Association, which is funded by Cargill, Archer Daniels Midland, Bayer, Monsanto, and Pioneer Hi-Bred. "Issues such as protecting food security and poor farmers' livelihoods are not on its agenda" (ActionAid International 2006, 30).

The United States and the European Union (EU) have been able to exploit resource disparities and shift forums whenever it suits their interests. This holds true for the horizontal shift from the WIPO to the WTO in 1986, when the United States first brought intellectual property into the multilateral trade forum, as well as the vertical shifting between multilateral, regional, and bilateral negotiations. At the end of the Uruguay Round of trade talks, negotiators did not share consensual assessments of the Agreement on Trade-Related Intellectual Property Rights (TRIPS) in the WTO. The United States and the European Union saw TRIPS as a floor—a minimum baseline for intellectual property protection. By contrast, developing-country negotiators saw it more as a ceiling—a maximum standard of protection beyond which they were unwilling to go. These divergent perceptions foreshadowed the continuing controversies that shape the politics of agricultural biotechnology.

With the advent of genetic engineering, plant breeders sought even stronger protection of their investments than afforded through PBRs by pursuing patent protection. TRIPS Article 27 permits the exclusion of plants and animals from patentability, but 27.3(b) requires that members provide protection for plant varieties or an "effective *sui generis*" (uniquely tailored) system. However, there really is no consensus on what a sui generis system needs to include and the negotiation history of Article 27 provides little guidance (Sutherland 1998, 295).

Agrifood corporations and their representatives, such as CropLife International, have been pushing the Union for the Protection of New Varieties of Plants (UPOV) as the model sui generis system. UPOV, to which sixty-three (UPOV 2006) mainly industrialized countries subscribe, was last amended in 1991. Before 1991, UPOV provided three limitations on plant breeders' monopoly rights. First, other breeders could freely use UPOV-protected varieties for research purposes. Second, farmers could reuse the seed for the following year's harvest under certain conditions. Third, plant breeders were forced to choose to protect their plant varieties with either a Plant Breeders' Right *or* a patent. The 1991 revision (UPOV91) "narrowed down the exemption for competing breeders and it deleted the so-called farmer's privilege" (GRAIN 1999, 2). Farmers have faced increasingly restrictive rules: "Although the UPOV system allows on-farm replanting, its rules restrict farmers' freedom to buy seed from sources other than the original breeders" (UNCTAD-ICTSD 2003, 107). UPOV91 "does not authorize farmers to sell or exchange seeds with other farmers for propagating purposes"

(Helfer 2002, 17). UPOV91 "permits member states to protect the same plant variety with both a breeders' right *and* a patent" (Helfer 2002, 15, emphasis added). Any country wishing to join UPOV today must sign the 1991 treaty, which is very generous to the corporate plant breeder. Indeed, UPOV was designed for industrialized, large-scale, agricultural production, not subsistence or smallholder farming (Rajotte 2008, 158). While corporate plant breeders would prefer that developing countries adopt UPOV91 as their domestic legislative standard, these countries are by no means required to do so.

In developing countries a large number of farmers are smallholders, who do not participate in the market in any substantial way. Indeed, in developing countries "80 percent of the total seed supply is produced on farm" (van Wijk 2004, 121). Smallholder farmers engage in seed saving, replanting, and "across-the-fence" exchange. This is especially the case for countries (many in Africa) where neither the public nor the private sector plays a significant role in producing and distributing seed (UNCTAD-ICTSD 2003, 107). The smallholder farming sector plays an important role in contributing to national food needs: "Small holder farmers produce fifty-one percent of Latin America's maize, seventy seven percent of its beans and sixty one percent of its potatoes while in Africa they produce the majority of grains and legumes and almost all root, tuber, and plantain crops" (Lettington 2003, 13). For example, in Peru between 50 and 60 percent, and in Kenya between 70 and 80 percent of the population depends on smallholder agriculture for their livelihood (p. 13).

In a comparative study of smallholder farming in Peru and Kenya, Lettington found that plant variety protection (PVP) legislation failed "to create solutions to existing problems" (p. 32). It creates incentives that direct energy away from catering to subsistence farmers' needs in favor of large commercial agricultural enterprises. It also facilitates the promotion of the use of commercial seed (as opposed to landraces or "wild" cultivars). "The end result," Lettington says, "has been a hastening of the deterioration of food security in these areas" (p. 34). He suggests that governments seeking to limit the cost of seed in economically and climatically marginal areas may "need to place limits on the nature of intellectual property rights" (p. 34), and recommends that "the activities of smallholder farmers, in particular the saving, use, exchange, and sale of farm-saved seed, should be explicitly stated as not subject to the rights of intellectual property rights holders. . . . Limited exceptions to intellectual property rights should be permitted to promote the adaptation of protected products to the needs of smallholder farmers" (p. 8).

Countries eschewing the UPOV system can adopt sui generis systems of protection that allow "farmers to acquire PBR-protected seed from any source and/or [to] require protected varieties to display qualities that are genuinely superior to existing varieties" (UNCTAD-ICTSD 2003, 107). A number of developing countries have favored a sui generis approach that explicitly honors farmers' rights. TRIPS, unlike UPOV91, preserves the right of subsistence farmers to exchange seed. States are free to incorporate both subsistence-farmer exemptions and research exemptions in national plant breeders' rights legislation (Barton 2003, 23–24). In countries lacking significant private-sector competition, public-sector seed provision will be important to promote competition to stimulate both variety and lower prices (Barton 2003, 14).

At the behest of their life science firms, playing a multilevel, multiforum governance game, the United States and the European Union have been able to extract a high price from economically more vulnerable parties eager to gain access to large, affluent markets (Abbott 2005, 350–354; Correa 2004; Vivas-Eugui 2003). The global agriculture firms have an ambitious global regulatory agenda. Bilateral Investment Treaties, Bilateral Intellectual Property Agreements, and regional Free Trade Agreements concluded between the United States and developing countries, and between the European Union and developing countries, invariably have been TRIPS-Plus (Drahos 2001; Dutfield 2002a). TRIPS-Plus agreements require signatories to adopt standards of intellectual property protection that go well beyond their TRIPS obligations. In the intellectual property provisions covering agriculture, developing countries are most often required to ratify or accede to UPOV91 as their sui generis system of protection, and "to undertake 'all reasonable efforts' to make patent protection available for plants" (South Centre and CIEL 2004, 12). The provisions of the Central American Free Trade Agreement contain these very provisions (IFAC-3 2004, 13). According to industry advisors to the USTR, the United States seeks to have negotiating partners adopt the same levels of intellectual property protection as found in the United States (IFAC-3 2004, 12).

The United States and its industries increasingly are coordinating enforcement through a number of venues. Industry representation in the USTR advisory committees, overlapping memberships in industry associations, such as CropLife International, and ad hoc mobilization vehicles such as the pro-industry NGO American Bio-Industry Alliance (ABIA) increase the information exchange among private actors and the USTR to monitor compliance, negotiate and enforce TRIPS-Plus deals, and lobby at national and multilateral levels. This thick network has resulted in a centralized system of private governance that enlists the USTR for legitimation and enforcement and heightens opportunities for rent seeking (Drahos 2004, 77).

Focusing on the formal features of intellectual property law, texts, and institutions, one sees plenty of room for state discretion and flexibility in adapting the global minimum standards to local concerns. However, this formal universe is embedded in a system of asymmetrical power relationships that constrain weaker states' abilities to take advantage of the flexibilities crafted into the law. They are eager to attract foreign investment and worry about alienating potential foreign investors. They also seek access to technologies that may help them to further develop, provide reliable nutrition, compete in global markets, and address a myriad of pressing problems. A number of nontrade motivations, including geopolitics and security, may also spur agreement to suboptimal economic deals with stronger countries (Drahos and Tansey 2008, 200). Most of these countries lack significant bargaining leverage to resist the high-pressure tactics of the USTR and the industries it represents.

Discursive Dimensions of Agriculture: Arguments and Actors

While agriculture for development and access to seeds is a complex topic, some discursive frames are particularly confusing. Terms like "food security" mean different things to different people. To some, it means reliance on landraces, local cultivars, organic farming methods, and the preservation of biodiversity. To others, it means higher yield, pest-resistant, drought-tolerant, saline-tolerant, and/or stress-tolerant crops via genetic modification (Williams, chapter 6, this volume). The focus on global biotechnology firms necessarily involves discussion of genetically modified organisms (GMOs). Fundamentally, there are three frames around GMOs. I will discuss each of these in turn.

Some maintain that GMOs are a curse for rich and poor alike. They argue that they damage ecosystems and upset ecological balances. They argue that GMOs reduce biodiversity and that windborne seeds can pollute non-GMO land. GMO technology promotes monocropping and encourages large-scale farming for export in lieu of staple crops.

Anti-GMO activists, such as Greenpeace and Oxfam, support organic farming, low-input farming, the use of landraces, on-farm saved seeds, and biological insecticides (Pistorius and van Wijk 1999, 194). Others, especially in Europe, highlight the potential health risks of using GMOs (Falkner, chapter 8, this volume).

Other anti-GMO arguments emphasize more specifically the dangerous impacts of adopting GMO technologies on poor farmers, who are ensnared in inextricable webs of dependency on multinational corporations for expensive inputs. Indian activist Vandana Shiva has dramatized this antiglobalization strand with her accounts of farmer suicides in Andhra Pradesh and Maharashtra (Shiva 2001; but see Herring 2005 and Ramanna 2006). Many African countries are appreciably wary, on political economy grounds, of GMO pollution of their export crops. Those countries exporting to Europe would suffer market loss because Europe does not accept GMO agricultural products (Clapp 2006; but see Paarlberg 2001).

On the other hand, some see GMOs as a blessing. Agrifood corporations promote GMOs as technologies that can increase yields, improve productivity, and solve the problem of hunger. The poor benefit from agricultural technology where "it directly produces more food for poor small landholders; . . . where adoption of high-yielding varieties . . . increases the demand for labour . . . and provides income opportunities for poor farmers; and . . . where increased productivity leads to lower food prices, particularly for staples" (Thomas 2005, 78). To the extent that GMOs serve these goals, they could have a positive impact on alleviating poverty and malnutrition. GMOs, such as *Bt* cotton, can render plants toxic to prominent pests, reduce crop losses, and reduce the need for toxic pesticides. They can also contribute to a country's competitiveness in globally integrated markets in agricultural products (Fukuda-Parr 2007).

These pro-GMO and anti-GMO discourses do not strictly fall along "industrialized country–Global South" or "industry–public sector" fault lines. Fukuda-Parr's (2007) research presents a far more variegated picture. A third narrative highlights the role of GMOs as "farmers' choice" (Ramanna 2006, 10). Based on the marketing and adoption of illegal *Bt* cotton seeds in Gujarat, India, some recast the GMO debate as one of farmers' choice. Pro-GMO advocates highlighted that the anti-GMO campaigns, such as Shiva's, were out of touch with the actual farmers that they purported to represent. When Monsanto sought to pressure the government of Gujarat to burn the illegal plantations, farmers who had planted the illegal Bt seeds protested so vehemently that the government backed off (Ramanna 2006, 10; Herring 2006). The farmers'-choice rationale depicts farmers not as victims of globalization but rather, "as decision makers and voters for the technology" (Ramanna 2006, 11). GMO uptake on a grand scale by China, Argentina, and Brazil underscores this element of choice (Fukuda-Parr 2007; Newell, chapter 9, this volume). China is moving full speed ahead in public-sector agricultural biotechnology and can hardly be depicted as a victim of Monsanto.[2]

Overall, the discursive battle rages on. No *one* compelling frame has yet emerged. Farmers' rights emerged as a frame to counter plant breeders' rights, yet has not triumphed. The latest discursive move of TNCs is to portray proprietary genetic modification as the silver bullet to ensure food security in the face of climate change (ETC Group 2008a, 2008b). NGOs will be quick to counter this frame, highlighting the importance of biodiversity, the precautionary principle, and the dangers of cartels. While competing discourses can create opportunities for mobilization and coalition building, or what Drahos and Tansey (2008, 201) refer to as "floating points of leverage," if no *one* discourse emerges as more persuasive or decisive, then instrumental and structural power are more likely to carry the day.

Given that some developing countries and, yes, some subsistence farmers seek access to GMO technology, the salient political question is: What obstacles do they face in achieving that goal? In examining the regulatory environment, the central questions are: How much discretion do states have in limiting intellectual property rights to support smallholder agriculture and environmentally sustainable agricultural practices, and to obtain access to GMO technology?

Some analysts argue that a focus on intellectual property rights may be misplaced. Most biotechnology is not patented in developing countries; therefore, freedom to operate is not greatly curtailed (Binenbaum et al. 2003). Developing countries suffer from lack of adequate public-sector investment and core capacities. The high costs of testing new products for efficacy and safety pose another important barrier to public and nonprofit agricultural researchers in developing countries (Fukuda-Parr 2007, 230). Furthermore, seed patents are hard to enforce, as evidenced in the widespread adoption of "illegal" *Bt* cotton seed in Gujarat, India, and the spread of unauthorized GMO seeds from Argentina to

Brazil (Herring 2005; Smolders 2005; Newell, chapter 9, this volume). Brazilian and Indian farmers eagerly adopted GMO soy and cotton seeds (respectively) on a massive scale, demonstrating that transgenic technology dissemination "can rapidly occur without intellectual property protection and without sponsorship by public authorities or large firms" (Wright and Pardey 2006a). Furthermore, the World Trade Organization's Agreement on TRIPS affords farmers substantial flexibilities. So why focus on intellectual property rights?

How Intellectual Property Protection Can Be an Obstacle to Access

Intellectual property rights have been dramatically expanded in recent years to cover things such as data, genes, entire plant species, and local agricultural practices (Arup 1998, 367–381; Correa 2002, 550). The consequences include enhanced political power of agrifood corporations, a reduction in the number of suppliers of certain kinds of technology, a reduction in competition, and higher costs of technology. Some suggest that excessively strong property rights over plant biotechnologies restrict access to research results "in ways that curtail the freedom to operate for research conducted in or on behalf of poor countries, to the detriment of developing-country food-security prospects" (Pardey, Koo, and Nottenburg 2003, 9).

The market-based justification for strong intellectual property rights is that patents and licenses provide incentives to "increase the number of commercially available products and thereby serve the public interest" (Lieberwitz 2004, 782). Yet, which publics are served? In agriculture, stakeholders include private-sector seed companies, public corporations, research institutes, and resource-poor farmers. Rights holders benefit, as do those who have the resources to participate in the commercial market. But market-based solutions alone fail to serve the poor, such as smallholder subsistence farmers. Lettington (2003, 7) observes that "approximately 75 percent of the world's undernourished are smallholder farmers." Furthermore, about 75 percent of Sub-Saharan Africa's people are rural and depend on agriculture for their livelihoods (Taylor and Cayford 2004, 280).

The power of patents is the power to *withhold*. Frequently rights holders treat patent portfolios as strategic business assets. Increasingly, patent holders are opting not to grant commercialization licenses; licensing to entities that will not grant others a license; seeking unreasonable terms for commercial licenses; and blocking applications of their technologies by competitors and acquisition targets (Binenbaum et. al. 2003, 314). As discussed earlier, climate change–related genes are just the latest target of this practice of strategic patenting. According to Richard Jefferson, founder and chief executive of Cambia, an Australia-based nonprofit institute that helps to reduce the transaction costs of acquiring gene tools and endorses open-source approaches to biotechnology, with contemporary gene-patenting practices "the little guys shake out and the big guys end up in a place a lot like a cartel" (quoted in Weiss 2008, A04). Wright and Pardey (2006a, 103) have documented numerous examples of this and survey research that highlights the constraining effects of intellectual property rights on agricultural public research institutions charged with "seeing research through to commercialization."

Patent thickets have proliferated, in which overlapping patent rights require those seeking to commercialize new technology to obtain licenses from multiple patent holders (Carrier 2003, 1090–1091; Rai and Eisenberg 2003). Krattiger (2004, 21) points out that few intellectual property exchanges "are complete enough to allow a prospective licensee to assemble all the needed licenses to obtain freedom-to-operate." This raises entry costs for prospective competitors. When genomics researchers Peter Boyer and Ingo Potrykus sought to make their vitamin A–enriched golden rice available to the International Rice Research Institute (IRRI), they faced significant delays because the technology was divided among seventy patents held by thirty-two organizations (Phillips 2004, 181). The transaction costs delayed access for over a year and a half. Wright and Pardey (2006a, 104) found that in a number of cases, "Freedom to operate was . . . a serious barrier to a system of nonprofit innovation that has responsibility for development to the point where they [innovative products and techniques] were made available to farmers in the field."

Breeders build on existing techniques and varieties to develop new transgenic plant varieties. The breeder of a new variety may face a bundle of property rights such as "plant variety rights, patents on plants, as well as several patents relating to transformation technology, the selectable marker employed, the gene coding for the protein, the promoter, and various regulatory elements and modifications needed to express genes adequately in plant cells" (Pistorius and van Wijk 1999, 148). This technology fragmentation is an increasing problem; if a rights holder denies just *one* element of this package of technologies, commercialization of the new variety could be blocked (Krattiger 2004, 31; Pistorius and van Wijk 1999, 148). Wright and Pardey (2006b, 21) argue that with the proliferation of biotechnology patents, "In addition to the aggregate value of the rent transfers to prior patent holders, the costs of actually consummating licensing deals can be very significant." Private intellectual property rights could block public-sector access to "innovations and germplasm that may be adaptable to smallholder needs . . . while also limiting public research options" (Lettington 2003, 8).

Even though developing countries are just now starting to implement stronger protections for agricultural biotechnology, some firms have deployed bogus US patents to delay or block commercialization of developing countries' agricultural products. In one heinous instance, a vice president of Del Monte Fresh Produce wrote a Central American researcher to warn him against working on Del Monte Fresh Produce pineapple plant material, stating that Del Monte owned a US Plant Patent. Wright and Pardey (2006a, 106) state that "the variety in question was not patented, and . . . it had been refused a US patent in 1992." This deliberate and strategic obfuscation deters developing-country researchers from moving forward. Wright and Pardey (2006a, 106) maintain that "the cost of challenging the validity of such [bogus] patents is likely to place such a heavy burden on less-developed country researchers and producers that competition in major markets might be eliminated, or delayed for years."

In the agrichemical and biotechnology sector, global firms are pressing for a new form of intellectual property right that would effectively extend the monopoly privileges beyond the life of the patent. Under Article 39.3 of TRIPS, WTO members must protect undisclosed test data on agricultural biotechnology and chemical products against unfair competition. Agrichemical firms are required to submit safety and efficacy test data as part of the regulatory approval process. However, various bilateral and regional Free Trade Agreement (FTA) provisions require signatories to grant at least ten years of exclusivity for agrichemical test data, counted from the date on which the product was approved, regardless of whether it was patented (Correa 2006, 401). These provisions require generic competitors to rely on their own field test data, rather than relying on the safety and efficacy findings of the brand-name biotechnology in the regulatory approval process. By rejecting bioequivalence, these companies have acquired a new form of intellectual property right in their test data and information generated by that data (Shadlen 2005, 19). Agrichemical and biotechnology firms favor data-exclusivity provisions because they offer new rights and opportunities to maximize returns on their products by delaying competition.

While public land grant universities conducted extensive agricultural research that they made freely available, this was changed with the 1980 Bayh-Dole Act, which allowed "grantees to seek patent rights in government-sponsored research results" (Rai and Eisenberg 2003, 290). Policymakers presumed that many inventions with commercial potential lay fallow in university laboratories, and that patenting opportunities would give universities incentives to "scour" research labs for significant and marketable inventions (Lieberwitz 2004). The Bayh-Dole Act has prompted a flurry of university patenting activity, at least a tenfold increase since 1979 (Rai and Eisenberg 2003, 292). University patent portfolios help to attract private-sector funding, especially in biotechnology; thus, the Bayh-Dole Act has had beneficial effects in generating revenue for cash-strapped public universities.

The Bayh-Dole Act has created new divisions within universities. The legislation makes no distinction between upstream and downstream research, and as a result more and more research tools have become patent-protected (Rai and Eisenberg 2003, 290–291), dramatically reducing open access to research tools. Technology transfer offices are charged with patenting and licensing technology to generate revenue for the institution. Research scientists are more interested in having access to "open science" (Rai and Eisenberg 2003, 305). Bayh-Dole has also increased university collaboration with private-sector biotechnology firms, and has raised many questions about academic freedom, research priorities, and incentives. Some critics have asserted that universities have lost their sense of public mission (Lieberwitz 2004).

Access-blocking technologies are proliferating rapidly in the current environment. In 1998 Delta and Pine Land, a subsidiary of Monsanto, along with the US Department of Agriculture (USDA), obtained a patent on a copy-protection technique for plants—Genetic Use Restriction Technologies (GURTs), aka "Terminator" technologies, which block the plant from replicating itself by rendering harvested seeds sterile. It is a strategic business technology designed to "prevent farmers from replanting saved seed and thereby undercut seed company monopolies" (Dutfield 2003b, 491). Syngenta, DuPont, and BASF were granted or applied for patents for GURTs in 2003 (van Wijk 2004, 122), and

Syngenta filed a GURT patent in 2005 and owns over half of the GURT patents issued so far (Shi 2006, 10–12). Both the USDA and Delta and Pine Land seek to expand the use of GURTs to increase the proprietary value of US-owned seed companies and to open developing-country markets. According to Harry Collins of Delta and Pine Land, co-owner of the patent with the USDA, the patent "'has the prospect of opening significant worldwide seed markets to the sale of transgenic technology for crops in which seed currently is saved and used in subsequent plantings'" (quoted in Dutfield 2003b, 493).

Forum Shifting: WIPO, CBD, and FAO

The WTO is the most important institution governing global intellectual property policy. It embodies hard law, notably TRIPS, which is binding and enforceable. However, the WIPO, the CBD, and the FAO are all actively engaged in making public international law in intellectual property as well, albeit soft law (neither binding nor enforceable). The following section discusses forum shifting and maps out the broader institutional terrain. Overall, agrifood corporations and their supportive governments have more resources to keep abreast of policy developments in these multiple forums, despite the fact that information technologies have made global mobilization for NGOs and developing countries that much easier.

Governments, private actors, and NGOs all engage in forum shifting. They can shift forums horizontally across institutions. They can shift forums vertically, such as the United States' recent efforts to use bilateral and regional free trade agreements to secure "TRIPS-Plus" protection in developing countries (Drahos 2001, 792–793).

Participants usually engage in forum shifting to effect change. According to Helfer (2004, 53), there are four main reasons why actors choose to shift forums: "to help achieve desired policy outcomes, to relieve political pressure for lawmaking in other international venues to generate counterregime norms, and to integrate those norms into the WTO and WIPO." Only the second, the "safety valve" strategy, is pursued in order to preserve the status quo. Forum shifting can act as a safety valve, when "consigning an issue area to a venue where consequential outcomes and meaningful rule development are unlikely to occur" (Helfer 2004, 56).[3]

Actors may choose among different institutions, favoring those that afford them better access or those whose philosophies resonate more closely with their own goals. This can provide them with opportunities to propose and experiment with policy approaches (Helfer 2004, 55). Forum shifting can provide governments with a "safe space" in which to exchange information, develop soft law, and craft viable policy alternatives that address their concerns (Helfer 2004, 58). For example, soft law forums like the World Health Organization and the CBD have become significant incubators of alternative approaches to intellectual property protection.

This strategy can also support the development of competing discourses that can change the way parties read TRIPS and are willing to apply it.[4] Furthermore, these competing discourses can challenge various domestic political bargains and integrate a broader range of viewpoints and parties into the issues. They can raise the political costs of defending the status quo (Sell and Prakash 2004, 143–175). The following discussion surveys TRIPS-related activity in diverse forums.

World Intellectual Property Organization

In the mid-1980s dissatisfied American negotiators shifted intellectual property deliberations out of WIPO and into the GATT. Americans, seeking high protectionist norms for intellectual property, favored GATT because there they could link intellectual property protection to trade. The US negotiators anticipated better results owing to the large and attractive US market that could be used as negotiating leverage. The forum shifting of the mid-1980s was a major blow to WIPO's morale and prestige. However, WIPO has bounced back with renewed energy. Since then, WIPO has transformed itself into a more entrepreneurial agency, with a mission to prove its continued relevance to intellectual property owners.

Unlike many international organizations, WIPO is almost self-sufficient. Rather than relying on government handouts, WIPO earns nearly 90 percent of its operating budget from its administration of the Patent Cooperation Treaty (PCT) (Doern 1999, 111; WIPO 2001). The biggest users of the PCT are the global corporations that produce knowledge-intensive products and processes, including the global life science industries. These corporations also are the most ardent champions of high protectionist norms for intellectual property. Most observers regard WIPO as being very sympathetic to industry interests. Furthermore, developing-country WIPO delegates come from intellectual property offices that are socialized to favor protection (Drahos 2002, 785);

the professional camaraderie among intellectual property lawyers from both North and South means that discussions are far less adversarial than in WTO.

WIPO provides technical assistance to developing countries as they seek to comply with TRIPS. The former head of WIPO, Kamil Idris, published a booklet in 2004 titled *IP [Intellectual Property] as a Power Tool for Development* that reflected industry perspectives endorsing high standards of intellectual property protection. Those who pay WIPO's freight undoubtedly shape its advocacy, and it has urged a number of developing countries to adopt TRIPS-Plus provisions in their national legislation (Deere 2009).

By the late 1990s intellectual property issues had received extensive critical attention in the WTO. In 1999 the United States and the European Union shifted forums to take their unmet concerns to WIPO and to pursue a Substantive Patent Law Treaty (SPLT) to harmonize patent law globally (Musungu 2005). WIPO had been conducting negotiations on an SPLT, and many observers suspected that TNCs pushed for this forum shift to pursue a TRIPS-Plus agenda that had become stymied at WTO. The biotechnology industry eagerly seeks to move to a uniform global higher-standard patent system.

Developing countries also seized opportunities to press their agendas, which are more fully developed in other venues, *within* WIPO. They sought to link biodiversity issues to SPLT negotiations by proposing the incorporation of the CBD recommendation that intellectual property applicants, when using genetic resources, prove that they had obtained prior informed consent to access those resources (Helfer 2004, 62). In response, WIPO agreed to establish a separate body within WIPO to address intellectual property aspects of resources and traditional knowledge (South Centre and CIEL 2005, 16). Subsequently, the Intergovernmental Committee on Intellectual Property, Genetic Resources, Traditional Knowledge and Folklore (IGC) conducted a number of studies reflecting the developing countries' concerns as reflected in the CBD. Developing countries have also requested that the WIPO Secretariat examine the implications of the SPLT for the IGC's work, illustrating "their increasing recognition of the need to coordinate lawmaking not only across different . . . venues . . . but also in different fora within the same intergovernmental organization" (Helfer 2004, 71).

In 2004 a group of developing countries (Group of Friends of Development) proposed a development agenda for WIPO.[5] The United States objected, arguing that WIPO is not a development agency but an organization specializing in intellectual property (Musungu 2005, 4; WIPO 2005). Those favoring high standards of intellectual property protection, such as the United States, sought to confine discussions of "development" issues such as disclosure of origin and genetic resources to the IGC (presumably where "words don't matter").

In June 2005 the Friends of Development, led by Brazil, refused to discuss an SPLT if there was no movement forward on the development agenda (ICTSD 2005, 22). The group succeeded in halting progress on an SPLT, holding it hostage to meaningful, substantive progress on a development agenda.

The life science conglomerates resist any attempt at being forced to include disclosure of origin in patent applications. They argue that it would inject unacceptable levels of uncertainty into the process of commercializing biotechnology. They claim that this uncertainty would have a chilling effect on investment in bioprospecting (Gorlin 2006, 4). Industry views these agriculture issues as investment issues. Since the 1980s when it linked intellectual property to trade, it underscored a strong relationship between high levels of intellectual property protection and foreign investment. Given that many developing countries seek foreign investment, industry leverages this vulnerability to press its agenda (Correa 2004, 345–346).

WIPO is institutionally constrained from pressing very far on issues that its main constituency (users of the PCT) opposes. WIPO has worked hard to get back into the good graces of the industrialized countries since 1986, and it seems unlikely that it would jeopardize its position by directly challenging the highly protectionist agenda (May and Sell 2005, 214). Playing on this very sensitivity, the US delegate warned WIPO not to pursue a substantive development agenda (IP-Watch 2005a). A number of industry representatives, including former US PTO official Bruce Lehman, urged WIPO not to be diverted from its mission by a "so-called development agenda" (IP-Watch 2005b). In June 2005 the US PTO bemoaned the SPLT impasse, stating that "the impasse also raises serious questions as to whether WIPO is even a viable forum for further meaningful patent discussions" (PTO 2005). At the November 2005 WIPO General Assembly, industry lobbyists showed up to express their opposition to a development agenda; "one industry source said they were there for an 'anti-development agenda'" (IP-Watch 2005c).

Nonetheless, WIPO held the first of two week-long meetings of the Provisional Committee on Proposals Related to a WIPO Development Agenda in late February 2006. The United States is determined not to let WIPO's mandate to promote intellectual property protection be compromised by the development agenda. Industry was out in force for these meetings. Director General of Brussels-based CropLife (a biotechnology and crop industry association), Christian Verschueren, expressed concern that " 'the entire IP concept is under attack' " and that " 'the industry is being put into a defensive corner . . . and the need for IP is no longer taken for granted' " (IP-Watch 2006a).

At a side event that CropLife International hosted during the WIPO meetings, Deputy Director General of WIPO Rita Hayes underscored the positive role that intellectual property protection plays in agriculture. She highlighted golden rice as "an example of a product that was IP-protected but licensed and available in a number of developing countries" (IP-Watch, 2006b). Recall that patent thickets actually delayed the availability of golden rice for over a year and a half (Phillips 2004, 181).

Convention on Biological Diversity

The CBD emerged out of the 1992 United Nations Environment Program's "Earth Summit" in 1992. It was a response to growing alarm about species loss, the erosion of genetic diversity, and the accelerating destruction of rainforests. Unlike TRIPS and WIPO, CBD more explicitly incorporates intellectual property provisions that developing countries favor. Article 8(j) recognizes communal or traditional knowledge. The CBD conception challenges the TRIPS view that endorses the Western, individualistic conception of knowledge ownership; this Western perspective draws a sharp line between "folklore" and "science." The CBD stresses that biological resources are the sovereign resources of states, whereas TRIPS enforces private property rights over them. Article 8(j) seeks to promote the wider application of traditional knowledge, "with the approval and involvement of the holders of such knowledge" (CBD 1992). Many developing countries and NGOs endorse the CBD as a way of combating biopiracy, in which global life science corporations expropriate genetic resources and traditional knowledge without authorization or compensation. Article 8(j) calls for respect and preservation for "innovations and practices of indigenous and local communities embodying traditional lifestyles relevant for the conservation and sustainable use of biological diversity" (CBD 1992).

The CBD's Conference of the Parties (COP)—the CBD member states that decide how to apply and implement the CBD—has addressed its compatibility with TRIPS: "After the entry into force of TRIPS, developing states led by China and the Group of 77 and sympathetic NGOs such as the World Wildlife Fund began to express concern over the relationship between intellectual property rights and the CBD's access and benefit sharing rules" (Helfer 2004, 33). The COP adopted the Bonn Guidelines in 2002, which stipulated that applicants for intellectual property rights should disclose the origin of any genetic resources or related knowledge relevant to the subject matter. Such disclosures are meant to facilitate the monitoring of whether applicants have received prior informed consent of the country of origin and complied with the country's conditions of access (Helfer 2004, 29). The biotechnology industry adamantly opposes such provisions. Jacques Gorlin, speaking on behalf of the ABIA, stated that "we think industry's comparative advantage is telling a story. . . . The message will be: 'We are the people who you are depending on to generate commercial benefits. If you do it through the disclosure system, it ain't gonna happen' " (IP-Watch 2006b). While the CBD member states have urged cooperation with the WTO and WIPO, they have "pointedly refrained from ceding jurisdiction over biodiversity-related intellectual property issues to these organizations" (Helfer 2004, 34).

India was the first to call for the primacy of the CBD over TRIPS Article 27.3(b) (covering plant variety protection) and argued that TRIPS should be amended to comply with the CBD (Tejera 1999, 981). Participants negotiating TRIPS agreed to revisit Article 27.3(b) four years after the date of entry into force, which was in 1999. Since 1998 the TRIPS Council has engaged in discussions about the relationship between the CBD and TRIPS Article 27.3(b). The Article 27.3(b) review process bolstered developing-country negotiators' confidence and issue-specific knowledge (Hepburn 2002). These discussions have broadened to include acknowledgment of diverse sui generis approaches to plant-variety protection and concerns about indigenous peoples (Hepburn 2002). India, Brazil, and the African group submitted papers to the TRIPS Council and delegations also began to acknowledge the relevance of the CBD (Hepburn 2002). Developing countries succeeded in incorporating these issues into the formal negotiating mandate of the WTO's Doha Ministerial Declaration at Qatar in 2001 (Abbott 2002, 489). This constituted a frank recognition of conflicts that must be addressed. It also demonstrates the migration of an issue developed in the CBD into the WTO; the developing countries' proposals were derived from the Bonn Guidelines (Helfer 2004, 59–62). This is yet another instance of fomenting strategic inconsistency in order to bring conflicting values into sharp relief.

The TRIPS-CBD relationship has yet to be resolved. Developing countries seek to amend TRIPS so that it will support the objectives of the CBD (South Centre and CIEL, 2005, 1). Seventeen countries formed the Group of Like-Minded Mega-diverse Countries, representing between 60 and 70 percent of the biodiversity of the planet (South Centre and CIEL 2005, 2).[6] So-called megadiverse countries want to halt unauthorized and uncompensated commercialization of their biological and traditional knowledge resources. While states are free to adopt disclosure of origin, prior informed consent, and benefit-sharing policies in national patent systems, developing countries have argued that without an *international* agreement, national-level laws would do little to stop biopiracy (South Centre and CIEL 2005, 8). By contrast, the United States prefers contract law to more global regulation. Industry staunchly opposes any linkage between patents and benefit sharing, and advocates a contractual approach (ABIA 2005; Checkbiotech 2006).

While implementing a disclosure requirement would not stop all misappropriations of biological and traditional knowledge resources, it could help to improve the patent examination process regarding prior art. A September 2004 submission from Brazil, India, Pakistan, Peru, Thailand, and Venezuela highlighted that disclosure requirements would also "be useful in cases relating to challenges to patent grants or disputes on inventorship or entitlement to a claimed invention as well as infringement cases" (WTO et al. 2004, para. 5). This would help reduce the prevalence of patents that lack novelty or an inventive step.

Developing countries would prefer a binding requirement that contains noncompliance penalties through the WTO Dispute Settlement Mechanism, whereas the European Union has been sympathetic as long as "non-compliance with the requirement does not affect the validity or enforceability of the granted patent" (Correa 2005, 6). This would seem to undercut the purpose of curtailing biopiracy. Dutfield recommends requiring "proof of legal acquisition" to connect the patent system more strongly to the CBD's access and benefit-sharing provisions, especially for those countries directly providing the resources (Dutfield 2005, 2).

At the WTO's Hong Kong Ministerial meeting in December 2005, Brazil, India, and Peru aggressively negotiated to include provisions in the final Ministerial text exhorting the Director General to intensify consultations on outstanding implementation issues—specifically on the TRIPS-CBD relationship. India argued that without addressing the disclosure of origin of material in patent applications there would be no development package (IP-Watch 2005d). The United States and global pharmaceutical and biotechnology companies strongly resisted agreeing to negotiate the disclosure issue (IP-Watch 2005e).

Peru's strong position on disclosure at the WTO was at odds with its TRIPS-Plus bilateral Free Trade Agreement with the United States. The US-Peru bilateral FTA endorses industry's preferred contractual approach to genetic resources and benefit sharing. Some sources at the WTO meeting argued that Peru had sold out and feared that the FTA provisions meant that "companies could negotiate contracts with indigenous communities without any transparency and in this case without any requirements to disclose . . . from whom and where they obtained their resources" (IP Watch 2005e). The unequal bargaining power of the parties does not bode well for equitable deals. While Peru had been prominent in pushing for TRIPS-CBD linkage, critics blamed the United States for its strategy of dividing coalitions through bilateral deals. Thus by engaging in vertical forum shifting, the United States was able to undercut Peru's multilateral position.

In June 2007, developing countries gained broad support for a TRIPS amendment incorporating their desired CBD provisions. Over fifty (mostly developing) countries at the TRIPS Council meeting endorsed the proposal. Developed-country opponents argued that the issue should be discussed at World Intellectual Property Organizations in the IGC instead (IP-Watch 2007).

Food and Agriculture Organization

The FAO is charged with leading international efforts to defeat hunger. It focuses on developing rural areas, sharing policy expertise, and providing a neutral forum for states to negotiate agreements and debate policy (FAO 2008). In the early 1980s, "Cary Fowler, a former political activist who during the 1980s opposed the extension of IPRs to life-forms, decided that forum-shifting was a game that the weak could also play" (Dutfield 2002b, 14). According to Fowler (1994, 180), the progressive extension of property rights with no commitment to conserving genetic diversity led activists such as himself and Pat Mooney to "develop a strategy and set it to work in a new but potentially friendlier arena." The 1981 biennial conference of the FAO marked a shift in the discourse and strategy over plant genetic resources. Mooney and Fowler had consulted with the Mexican ambassador to the FAO prior to the meeting and had provided the Mexican president with reports on genetic resources. The ambassador arrived armed with proposals for the establishment of an international gene bank and an international legal convention for the exchange of genetic resources (Fowler 1994, 181).

Shifting to the FAO was a strategic move for two reasons: first, it shifted the center of gravity from American to developing-country interests, and second, it broadened the discourse beyond patenting plant varieties to the wider connections between patenting, genetic conservation, and development issues (Fowler 1994, 181–182).

At issue was the one-way flow of germplasm from the genetically rich geopolitical South to the North (GRAIN 2005b). During the United Nations debates over a New International Economic Order (NIEO) in the 1970s and 1980s, developing countries and industrialized countries clashed over what constituted the "common heritage of mankind" (Sell 1998, 88). Industrialized countries considered landraces, or wild cultivars, to be a common human heritage, ensuring free and easy access to germplasm from the South. However, they asserted property rights in industrialized countries' breeders' plant varieties. As Drahos and Tansey (2008, 203) point out, "In most fora . . . developed countries and business interests argue that one size does not fit all. When it comes to IP [intellectual property] they tend to argue that a minimum size fits all, with preferably an even bigger minimum. This is another example of sophistry." Raustiala and Victor (2003, 5) have referred to this distinction as "raw" versus "worked" germplasm. In the current discussions over traditional knowledge (TK), advocates for TK protection argue that what Northern plant breeders consider to be "raw" is in fact the product of on-farm innovations over generations.

In 1983 negotiators of the NIEO agreed to the International Undertaking on Plant Genetic Resources (IUPGR), against the adamant opposition of the United States, the United Kingdom, Australia, and major seed companies. The nonbinding undertaking was designed to ensure conservation and unrestricted availability and sustainable utilization of plant genetic resources for future generations by providing a flexible benefit and burden-sharing framework (Dutfield 2005, 15). Negotiators also established a new FAO Commission on Plant Genetic Resources where states could discuss and monitor the nonbinding undertaking. The United States tried to boycott the commission, but 93 countries were represented at its first meeting in 1985. It is now the largest FAO Commission with 161 members and the EU (Dutfield 2005, 15).

Beginning in 1989 the IUPGR incorporated Pat Mooney's notion of "farmers' rights" "to acknowledge the contribution that farmers have made to conserving and developing plant genetic resources" (Dutfield 2005, 15). The notion was to safeguard the rights of farmers to "work with, and live from, farming systems based on diversity, in the face of expanding monocultures and uniform seeds" (GRAIN 2005a). "Farmers' rights" was a deliberate counterframe to "plant breeders' rights." Developing countries sought to address the inequities arising from a system in which they lacked open access to improved varieties bred by seed companies, especially when some of those "worked" resources were based on developing countries' "raw" germplasm (Raustiala and Victor 2003, 12). The 1989 IUPGR asserted the principle of unrestricted access and common heritage of both raw and worked germplasm. But "the industrialized countries relied on the principle to continue their open access to raw PGR, yet refused to accept the undertaking's principle of open access to worked PGR" (Raustiala and Victor 2003, 13).

Developing countries took a new tack in the late 1980s and began to see property rights in raw germplasm, termed "green gold" (Kuanpoth 2005, 139), as a way to acquire wealth. Gene-rich countries could deny access to their biological riches and cut deals with companies that wanted to engage in bioprospecting. Over the years developing countries have been abandoning the "common heritage" conceptions of biological resources and have asserted their sovereign rights to their exploitation. Access can be privatized effectively under the rubric of sovereignty and attendant rights of states over their natural and cultural resources. Asserting property rights in biological resources has been controversial (for both worked and raw varieties). As Jack Kloppenburg has noted, one can question "whether the whole farmers' rights orientation was the proper way to go and whether there simply weren't too many contradictions embedded in trying to use the master's tools to dismantle the master's house" (quoted in GRAIN 2005b).

The 1991 Annex to the 1983 FAO Undertaking reflected this new approach and stated that "the concept of mankind's heritage is subject to the sovereignty of states over their plant genetic resources" (quoted in Raustiala and Victor 2003, 16). The 2001 FAO Treaty that came into force in 2004 has been sharply criticized as being a bonanza for plant breeders and a disaster for farmers' rights. On the plus side, first the FAO Treaty "recognizes that access itself is the main benefit to be shared, and aims to facilitate it rather than limit it by exclusive contracts and patents. Second, any monetary benefits generated through the system are to be pooled and used to support conservation and sustainable use efforts, rather than enrich any single provider" (GRAIN 2005a). It falls well short of a generalized system of mutual access to all plant genetic resources for food and agriculture. It clearly enshrines the notion of property rights over genetic resources. In the negotiations, developed countries resisted any measures that would prevent their patenting of genetic resources, whereas developing countries wanted to limit the scope of the treaty to protect any business opportunities that might result from providing individual genes on the global market (GRAIN 2005a).

In 2004 the FAO adopted a human rights and a food security frame for its "Voluntary Guidelines to Support the Progressive Realization of the Right to Adequate Food in the Context of National Food Security" (FAO 2004). 188 member states adopted the guidelines and pledged to work toward the goal of supporting the right to food. Advocacy groups supporting vulnerable populations endorse this approach because it focuses on the likely impact of trade policies and provides international mechanisms to hold economic actors accountable when domestic processes fail to promote or protect human rights (IATP 2005, 5). Under international law, a human rights rubric provides mechanisms for states, groups, and individuals to make their case (IATP 2005, 5).

Conclusion: The Way Forward

In recent years, public funding for applied agricultural research that is of direct use to farmers has shrunk. The ability to patent living organisms since *Diamond v. Chakrabaty* has enticed many private agrichemical firms "into plant breeding R & D, as they saw opportunities to dominate markets" (Baumuller and Tansey 2008, 186–187). The Bayh-Dole Act of 1980 reduced the easy access to university research, the mainstay of the land grant mission of American universities that used to give seeds out for free. Even the once–publicly funded Consultative Group on International Agricultural Research (CGIAR), faced with slashed budgets, has entered into partnerships with the Gates Foundation's (funded by

Microsoft) US$1.5 billion collaboration with BASF and Monsanto to develop stress-tolerant genes to respond to the challenge of climate change (Weiss 2008, A04).

With the overwhelming shift from public-sector to private-sector research and investment over the last several decades, it is easy to forget what a significant role public-sector research played in the development of modern agriculture. Innovation flourished through access to germplasm in the public domain. For example, between 1971 and 1990 the United States introduced 21.6 commercially successful new wheat varieties *per year*, worth about US$43 billion to the United States between 1970 and 1993 (Pardey, Koo, and Nottenburg 2003). Up to US$13.6 billion of that total benefit was attributable to varietal spillovers from the publicly funded International Maize and Wheat Improvement Center (CIMMYT) in El Batan, Mexico (Pardey et al. 2003, 15 at note 16). In maize and wheat, "Almost all the resulting improved varieties were made available without personal or corporate intellectual property rights" (Pardey, Koo, and Nottenburg 2003, 16). Between 1970 and 1993, the US economy gained between US$30 million and US$1 billion using rice varieties developed by the public-sector IRRI in the Philippines. Public-sector research has been profoundly important in agricultural development, and has been a most efficient and productive expenditure of public-sector monies.

Public land grant universities and their agricultural extension services have played a central role in innovation and dissemination (Schuh 2004, 359–371). Since the passage of the Bayh-Dole Act of 1980, American universities have been able to patent the fruits of federally funded research, creating incentives to assert proprietary rights over their innovations. They may feel caught between the conflicting imperatives of attracting private-sector funding and generating revenue through patenting activity on the one hand, and promoting access to agricultural biotechnology through humanitarian intellectual property policies on the other. Universities could play a significant role in promoting access to biotechnology suited for developing countries' needs. To this end, the Public Intellectual Property Resource for Agriculture (PIPRA) seeks to facilitate the development and dissemination of crops for developing countries and promote commons-based crop development (PIPRA n.d.). It is a public-sector initiative dealing with training, good practices, and the bundling of technologies. According to Krattiger (2004, 22), "The rationale for PIPRA is that IP is often unwittingly encumbered.

Universities . . . typically grant world wide exclusive licenses. Changing these licensing policies and retaining the rights for humanitarian uses in the developing world would make it much easier to transfer IP . . . from universities to the developing world."

Another promising nonprofit initiative is the Nairobi-based African Agricultural Technology Foundation (AATF). The UK Department for International Development, the Rockefeller Foundation, and USAID fund AATF, which "helps farmers—and African researchers—to access productivity-enhancing technologies held by the private sector that would otherwise not be available, owing to intellectual property rights" (Thomas 2005, 83). The AATF acts as a technology broker, and it "focuses on downstream activities, including the creation of local, national, and regional markets for the products produced from transferred technologies" (Krattiger 2004, 24). Critics worry that these arrangements may create long-term dependency or vulnerability. In any case, despite potential positive effects, they are no panacea.

While seeking to overhaul the current system is a tall order, Krattiger (2004, 16) suggests that it "should not preclude incremental changes in the system, such as open-source licensing, especially for research tools and processes." In biotechnology a number of projects are currently underway and are modeled after the open-source software movement. An Australian initiative, Biological Innovation for Open Society (BIOS), has adapted the free software movement to agricultural biotechnology to catalyze a new self-sustaining commons for researchers. BIOS aims to create portfolios of essential research tools and to license them under a license modeled after the GNU General Public License (GPL) (Kapczynski et al. 2005). Richard Jefferson, scientist and initiator of BIOS, stated that " 'so much of what we want to do is all tied up in somebody's intellectual property. . . . It's a complete sclerotic mess, where nobody has any freedom of movement. Everything that open source has been fighting in software is exactly where we find ourselves now with biotechnology' " (quoted in Kapczynski et al. 2005). Jefferson has created two technologies that circumvent proprietary tools for biotechnological crop improvement. One of these technologies, a method for introducing new genes into plants, bypasses numerous patents and has been successfully used in a Chinese project to create transgenic rice lines (Feldman 2004, 127).

In addition to these efforts to try to reclaim the commons and facilitate access, some have promoted the concepts of awarding prizes for innovation and of developing R&D treaties to require contributions toward the development of neglected public goods (Hubbard and Love 2004). In terms of domestic patent practices, raising nonobviousness standards (or instituting strict criteria for novelty and inventive steps), reducing the number of wrongly issued patents, and changing the presumption of validity would help to reduce patent abuses (Krattiger 2004, 15).

Returning to the notion of farmers' choice, GMO technology should be their *choice*. Farmers must be able to gain access to GMO technologies if they want them, whether for big export crops or smallholder staple foods. They also must be able to keep their fields free of GMOs if that is their preference. Addressing the needs of smallholder farmers in developing countries is imperative for a number of reasons. First, they contribute to the planet's biodiversity. Second, they innovate and develop staple crops suitable for their microclimates. This helps to promote both food security and biodiversity. Third, this sector employs millions of rural people. Any abrupt change to farming practices that would make these people lose their employment would have politically destabilizing effects that governments might not be able to manage. Since the public and nonprofit sector in developing countries will bear the lion's share of the task of agricultural innovation for staple crops, policies must be designed so that intellectual property rights do not prevent this sector from delivering essential goods.

Clearly, structural and instrumental powers play a huge role in shaping the politics of agricultural biotechnology and intellectual property. Discursive power plays some role; under the right conditions it can be transformative when used to activate "floating points of leverage" (Drahos and Tansey 2008, 209) that defy the outcomes expected based on structural and instrumental power. While forum shifting is a game that the weak can play, the strong have more resources to prevail in that endeavor. The dynamics of intellectual property policymaking in agricultural biotechnology suggest preponderant corporate structural and instrumental power, met with alternative discourses as well as technological "work-arounds" and open-source models that may render the property fences less high and less exclusionary. In conclusion, Geoff Tansey (2008, 220) notes that "it would be ironic—and potentially tragic—if just as other sectors are turning to and seeing the value of open source, informally networked means for innovation, farming and food, which has been based on such systems for millennia, moves in the opposite direction."

Notes

1. The Union for the Protection of New Varieties of Plants (UPOV) Convention, negotiated in 1961, established breeders' rights; the United States adopted the Plant Variety Protection Act in 1970 (Pistorious and van Wijk 1999, 84).

2. However, the case of Argentina is complicated by the fact that, unlike China, it is utterly dependent on foreign technology and adopted it in the context of IMF structural adjustment lending in the early 1990s (Chudnovsky 2007, 85).

3. This is the approach that the United States has been pursuing in discussions of a development agenda. The United States maintains that development issues do not belong in WIPO, and for the first several years recommended that the discussions be sidelined onto a committee with a mandate only to discuss the issue. Industry representatives have complained of WIPO's "mission creep."

4. E-mail from Gregory C. Shaffer, Professor of Law, University of Wisconsin Law School, to Susan K. Sell, November 23, 2003 (on file with author).

5. Argentina and Brazil presented the proposal to WIPO's General Assembly. The proposal was cosponsored by Bolivia, Cuba, the Dominican Republic, Ecuador, Egypt, Iran, Kenya, Peru, Sierra Leone, South Africa, Tanzania, and Venezuela (WIPO 2004).

6. Bolivia, Brazil, China, Colombia, Costa Rica, Democratic Republic of Congo, Ecuador, India, Indonesia, Kenya, Madagascar, Malaysia, Mexico, Peru, Philippines, South Africa, and Venezuela.

References

Abbott, Frederick. 2002. The Doha Declaration on the TRIPS Agreement and Public Health: Lighting a Dark Corner at the WTO. *Journal of International Economic Law* 5 (2): 469–505.

Abbott, Frederick. 2005. The WTO Medicines Decision: World Pharmaceutical Trade and the Protection of Public Health. *American Journal of International Law* 99 (2): 317–358.

ActionAid International. 2006. *Under the Influence: Exposing Undue Corporate Influence over Policy-Making at the World Trade Organization.* http://www.actionaid.org.

American BioIndustry Alliance (ABIA). 2005. Views on the Establishment of an ABS/IR. http://www.abialliance.com.

Aoki, Keith. 2003. Weeds, Seeds & Deeds: Recent Skirmishes in the Seed Wars. *Cardozo Journal of International and Comparative Law* 11 (2): 247–332.

Arup, Christopher. 1998. Competition over Competition Policy for International Trade and Intellectual Property. *Prometheus* 16 (3): 367–381.

Barton, John. 2003. Nutrition and Technology Transfer Policies. UNCTAD/ICTSD Capacity Building Project on Intellectual Property and Sustainable Development. August. http://www.iprsonline.

Baumuller, Heike, and Geoff Tansey. 2008. Responding to Change. In Geoff Tansey and Tasmin Rajotte, eds., *The Future Control of Food*, 171–196. London: Earthscan.

Binenbaum, Eran, Carol Nottenburg, Philip Pardey, Brian Wright, and Patricia Zambrano. 2003. South-North Trade, Intellectual Property Jurisdictions, and Freedom to Operate in Agricultural Research on Staple Crops. *Economic Development and Cultural Change* 51 (2): 309–335.

Carrier, Michael. 2003. Resolving the Patent-Antitrust Paradox through Tripartite Innovation. *Vanderbilt Law Review* 56 (4): 1047–1111.

Checkbiotech. 2006. Biotech Firms Form Alliance TRIPS. http://www.checkbiotech.org.

Chudnovsky, Daniel. 2007. Argentina: Adopting RR Soy, Economic Liberalization, Global Markets and Socio-Economic Consequences. In Sakiko Fukuda-Parr, ed., *The Gene Revolution: GM Crops and Unequal Development*, 85–103. London: Earthscan.

Clapp, Jennifer. 2006. Unplanned Exposure to Genetically Modified Organisms: Divergent Responses in the Global South. *Journal of Environment and Development* 15 (1): 3–21.

Convention on Biological Diversity. 1992. available at: http://www.cbd.int/convention/convention.shtml

Correa, Carlos. 2002. TRIPS and Access to Medicines: Internationalization of the Patent System and New Technologies. *Wisconsin International Law Journal* 20 (3): 523–550.

Correa, Carlos. 2004. Investment Protection in Bilateral and Free Trade Agreements: Implications for the Granting of Compulsory Licenses. *Michigan Journal of International Law* 26 (1): 331–353.

Correa, Carlos. 2005. *The Politics and Practicalities of a Disclosure of Origin Obligation.* Occasional Paper 16. Geneva: Quaker United Nations Office, January. http://www.geneva.quno.org.

Correa, Carlos Maria. 2006. Implications of Bilateral Free Trade Agreements on Access to Medicines. *Bulletin of the World Health Organization* 84 (5): 399–404. http://www.who.int/entity/bulletin/volumes/84/5/399.pdf.

Deere, Carolyn. 2009. *The Implementation Game: The TRIPS Agreement and the Global Politics of Intellectual Property Reform in Developing Countries.* Oxford: Oxford University Press.

Doern, Bruce. 1999. *Global Change and Intellectual Property Agencies.* New York: Routledge.

Drahos, Peter. 2001. BITS and BIPS: Bilateralism in Intellectual Property. *Journal of World Intellectual Property Law* 4 (6): 791–808.

Drahos, Peter. 2002. Developing Countries and Intellectual Property Standard-Setting. *Journal of World Intellectual Property* 5 (5): 765–789.

Drahos, Peter. 2004. Securing the Future of Intellectual Property: Intellectual Property Owners and Their Nodally Coordinated Enforcement Pyramid. *Case Western Reserve Journal of International Law* 36 (1): 53–78.

Drahos, Peter, and Geoff Tansey. 2008. Postcards from International Negotiations. In Geoff Tansey and Tasmin Rojette, eds., *The Future Control of Food*, 197–211. London: Earthscan.

Dutfield, Graham. 2002a. Sharing the Benefits of Biodiversity: Is There a Role for the Patent System? *Journal of World Intellectual Property* 5 (6): 899–932.

Dutfield, Graham. 2002b. *Trade, Intellectual Property and Biogenetic Resources: A Guide to the International Regulatory Landscape*. Background Paper for the Multi-stakeholder Dialogue on Trade, Intellectual Property and Biological and Genetic Resources in Asia. Geneva: ICTSD. http://www.ictsd.org/dlogue/2002-04-19/Dutfield.pdf.

Dutfield, Graham. 2003a. *Intellectual Property Rights and the Life Science Industries*. Aldershot, UK: Ashgate.

Dutfield, Graham. 2003b. Should We Terminate Terminator Technology? *European Intellectual Property Review* 25 (11): 491–495.

Dutfield, Graham. 2005. *Thinking Aloud on Disclosure of Origin*. Occasional Paper 18. Geneva: Quaker United Nations Office, October. http://www.quno.org.

ETC Group. 2008a. News Release: Gene Giants Grab "Climate Genes": Amid Global Food Crisis, Biotech Companies Are Exposed as Climate Change Profiteers. http://www.etcgroup.org/en/materials/publications.html?pub_id=688.

ETC Group. 2008b. Patenting the "Climate Genes" . . . And Capturing the Climate Agenda. *Communique* 99 (May/June). http://www.etcgroup.org/en/materials/publications.html?pub_id=687.

Feldman, Robin, 2004. The Open Source Biotechnology Movement: Is It Patent Misuse? *Minnesota Journal of Law, Science and Technology* 6 (1): 117–167.

Food and Agriculture Organization (FAO). 2004. Voluntary Guidelines to Support the Progressive Realization of the Right to Adequate Food in the Context of National Food Security. http://www.fao.org/righttofood.

Food and Agriculture Organization (FAO). 2008. FAO: About FAO. http://www.fao.org/about/about-fao.html.

Fowler, Cary. 1994. *Unnatural Selection: Technology, Politics and Plant Evolution*. Yverdon, Switzerland: Gordon and Breach.

Fukuda-Parr, Sakiko, ed. 2007. *The Gene Revolution: GM Crops and Unequal Development*. London: Earthscan.

Gorlin, Jacques. 2006. Additional Patent Disclosure Requirements for Biotech Inventions in Play in WTO, WIPO and CBD. *Stockholm Network: Know IP—Stockholm Network Monthly Bulletin on IPRs* 2 (1): 3–4.

GRAIN. 1999. UPOV on the War Path. *Seedling*, June. http://www.grain.org/seedling/?id=67.

GRAIN. 2005a. The FAO Seed Treaty: From Farmers' Rights to Breeders' Privileges. *Seedling*, October. http://www.grain.org/seedling/?id=411.

GRAIN. 2005b. Interview with Jack Kloppenburg. *Seedling*, October. http://www.grain.org/seedling/?id=414.

Helfer, Lawrence. 2002. *Intellectual Property Rights in Plant Varieties: An Overview with Options for National Governments*. FAO Legal Papers Online No. 31, July. http://www.fao.org/Legal/Prs-OL/lpo31.pdf.

Helfer, Lawrence. 2004. Regime Shifting: The TRIPs Agreement and New Dynamics of International Intellectual Property Lawmaking. *Yale Journal of International Law* 29 (1): 1–82.

Hepburn, Jonathan. 2002. *Negotiating Intellectual Property: Mandates and Options in the Doha Work Program*. Occasional Paper 10. Geneva: Quaker United Nations Office, November. http://www.quno.org.

Herring, Ron. 2005. Miracle Seeds, Suicide Seeds and the Poor: GMOs, NGOs. Farmers and the State in Social Movements. In Raka Ray and Mary Katzenstein, eds., *Social Movements in India: Poverty, Power, and Politics*, 203–232. New Delhi: Oxford University Press.

Herring, Ron. 2006. Why Did "Operation Cremate Monsanto" Fail? Science and Class in India's Great Terminator Technology Hoax. *Critical Asian Studies* 38 (4): 467–493.

Hubbard, Tim, and James Love. 2004. A New Trade Framework for Global Healthcare R&D. *PLoS Biology* 2 (2): 147–150.

Institute for Agriculture and Trade Policy (IATP). 2005. Planting the Rights Seed: A Human Rights Perspective on Agriculture and Trade in the WTO. 3D Trade, Human Rights and Equitable Economy Backgrounder No. 1 in the Thread series. March. http://www.3dthree.org/en/page.php?IDpage=38.

International Centre for Trade and Sustainable Development (ICTSD). 2005. WIPO Development Agenda Status Unclear. *Bridges* 9 (9): 22.

International Functional Advisory Committee on Intellectual Property Rights for Trade Policy Matters (IFAC-3). 2004. *The U.S.-Central American Free Trade Agreement (FTA): The Intellectual Property Provisions. Advisory Committee Report to the President, the Congress and the United States Trade Representative of the U.S.* March 12. http://www.ustr.gov/assets/Trade_Agreements/Regional/CAFTA/CAFTA_Reports/asset_upload_file571_5945.pdf.

IP-Watch. 2005a. Nations Clash on Future of WIPO Development Agenda. *Latest News*, April 11. http://www.ip-watch.org.

IP-Watch. 2005b. Non-Profits, Industry Offer Views on WIPO Development Agenda. *Latest News*, April 14. http://www.ip-watch.org.

IP-Watch. 2005c. Industry Concerned about Development Agenda at WIPO. *Latest News*, November 4. http://www.ip-watch.org.

IP-Watch. 2005d. India, Brazil Tie Biodiversity Negotiations to Doha Development Package. *Latest News*, December 15. http://www.ip-watch.org.

IP-Watch. 2005e. Peru Attempts Strong WTO Position on Disclosure Despite Weaker U.S. Deal. *Latest News*, December 17. http://www.ip-watch.org.

IP-Watch. 2006a. Agricultural IP Issues Debated at Industry Side Event to WIPO Meeting. *Latest News*, February 24. http://www.ip-watch.org.

IP-Watch. 2006b. Biotech Industry Fights Disclosure in Patents on Three IP Policy Fronts. *Latest News*, March 2. http://www.ip-watch.org.

IP-Watch. 2007. TRIPS Council: Big Boost for Biodiversity Amendment: Enforcement Debated. *Latest News*, June 6. http://www.ip-watch.org.

Kapczynski, Amy, Samantha Chaifetz, Zachary Katz, and Yochai Benkler. 2005. Addressing Global Health Inequities: An Open Licensing Approach for University Innovations. *Berkeley Technology Law Journal* 20 (2): 1031–1114.

Krattiger, Anatole. 2004. Financing the Bioindustry and Facilitating Biotechnology Transfer. *IP Strategy Today* 8. http://www.biodevelopments.org/ip/ipst8.pdf.

Kuanpoth, Jakkrit. 2005. Closing in on Biopiracy: Legal Dilemmas and Opportunities for the South. In Ricardo Melendez-Ortiz and Vicente Sanchez, eds., *Trading in Genes: Development Perspectives on Biotechnology, Trade and Sustainability*, 139–152. London: Earthscan.

Lettington, Robert. 2003. Small-Scale Agriculture and the Nutritional Safeguard under Article 8(1) of the Uruguay Round Agreement on Trade-Related Aspects of Intellectual Property Rights: Case Studies from Kenya and Peru. UNCTAD/ICTSD Capacity Building Project on Intellectual Property and Sustainable Development. November. http://www.iprsonline.org/unctadictsd/docs/lettingtonfinaldraft.pdf.

Lieberwitz, Risa. 2004. Book Review: The Marketing of Higher Education: *The Price of the University's Soul: Universities in the Marketplace: The Commercialization of Higher Education*. By Derek Bok. *Cornell Law Review* 89 (3): 763–800.

May, Christopher, and Susan K. Sell. 2005. *Intellectual Property Rights: A Critical History*. Boulder: Lynne Rienner.

Musungu, Sisule. 2005. *Rethinking Innovation, Development and Intellectual Property in the UN: WIPO and Beyond*. TRIPS Issues Papers 5. Ottawa, Canada: Quaker International Affairs Programme.

Paarlberg, Robert. 2001. *The Politics of Precaution: Genetically Modified Crops in Developing Countries*. Washington, DC: International Food Policy Research Institute.

Pardey, Philip, Bonwoo Koo, and Carol Nottenburg. 2003. Creating, Protecting, and Using Crop Technologies Worldwide in an Era of Intellectual Property. World Intellectual Property Organization, Geneva, WIPO-UPOV/SYM/03/4, October 10. www.wipo.org.

Patent and Trademark Office (PTO). 2005. Patent Law Harmonization Talks Stall; Brazil, Argentina, India Oppose Compromise. June 14. http://www.uspto.gov/main/homepagenews/bak2005jun14.htm.

Phillips, Ronald. 2004. Intellectual Property Rights for the Public Good: Obligations of U.S. Universities to Developing Countries. *Minnesota Journal of Law, Science & Technology* 6 (1): 177–185.

Pistorius, Robin, and Jeroen van Wijk. 1999. *The Exploitation of Plant Genetic Information: Political Strategies in Crop Development.* New York: CABI Publishing.

Pragnell, Michael. 2005. Agriculture, Business and Development. In Calestous Juma, ed., *Going for Growth: Science, Technology and Innovation in Africa*, 88–126. London: The Smith Institute.

Public Intellectual Property Resource for Agriculture (PIPRA). n.d. www.pipra.org.

Rai, Arti, and Rebecca Eisenberg. 2003. The Public Domain: Bayh-Dole Reform and the Progress of Biomedicine. *Law & Contemporary Problems* 66 (1): 289–314.

Rajotte, Tasmin. 2008. The Negotiations Web: Complex Connections. In Geoff Tansey and Tasmin Rojette, eds., *The Future Control of Food*, 141–167. London: Earthscan.

Ramanna, Anitha. 2006. *India's Policy on Genetically Modified Crops.* Asia Research Centre Working Paper 15. London: Asia Research Centre, London School of Economics. http://www.lse.ac.uk/collections/asiaResearchCentre/pdf/WorkingPaper/ARCWorkingPapers15AnithaRAMANNA2006.pdf.

Raustiala, Kal, and David Victor. 2003. *The Regime Complex for Plant Genetic Resources.* Working Paper 14, Program on Energy and Sustainable Development. Stanford, CA: Stanford University (on file with author).

Schuh, G. Edward. 2004. Intellectual Property Rights and the Land Grant Mission. *Minnesota Journal of Law, Science and Technology* 6 (1): 359–371.

Sell, Susan K. 1998. *North-South Politics of Intellectual Property and Antitrust.* Albany: State University of New York Press.

Sell, Susan K., and Aseem Prakash. 2004. Using Ideas Strategically: The Contest between Business and NGO Networks in Intellectual Property Rights. *International Studies Quarterly* 48 (1): 143–175.

Shadlen, Ken. 2005. *Policy Space for Development in the WTO and Beyond: The Case of Intellectual Property Rights.* Global Development and Environment Institute Working Paper No. 05–06. Medford, MA: Tufts University. http://ase.tufts.edu/gdae/Pubs/wp/05-06PolicySpace.pdf.

Shi, Guanming. 2006. "Intellectual Property Rights, Genetic Use Restriction Technologies (GURTS), and Strategic Behavior" Paper No. 156058 prepared for American Agricultural Economics Association Annual Meeting, Long Beach, CA. July 23–26. http://ageconsearch.umn.edu/bitstream/21434/1/sp06sh04.pdf

Shiva, Vandana. 2001. *Stolen Harvest: The Hijacking of the Global Food Supply.* London: Zed.

Smolders, Walter. 2005. Plant Genetic Resources for Food and Agriculture: Facilitated Access or Utility Patents on Plant Varieties? *IP Strategy Today* 13: 1–17. www.biodevelopments.org/ip/ipst13.pdf.

South Centre and CIEL (Center for International Economic Law). 2004. Intellectual Property and Development: Overview of Developments in Multilateral, Regional, and Bilateral Fora. *South Centre and CIEL IP Quarterly Update.* Second Quarter. http://www.ciel.org/Publications/IP_Update_2Q04.pdf.

South Centre and CIEL (Center for International Economic Law). 2005. Intellectual Property and Development: Overview of Developments in Multilateral, Regional, and Bilateral Fora. *South Centre and CIEL IP Quarterly Update.* Third Quarter. http://www.ciel.org/Publications/IP_Update_3Q05.pdf.

Sutherland, Johanna. 1998. TRIPS, Cultural Politics and Law Reform. *Prometheus* 16 (3): 291–303.

Tansey, Geoff. 2006. *Global Rules, Patent Power and Our Food Future: Controlling the Food System in the 21st Century.* IIIS Discussion Paper No. 130. Dublin: Institute for International Integration Studies, Trinity College, Dublin, April.

Tansey, Geoff. 2008. Global Rules, Local Needs. In Geoff Tansey and Tasmin Rojette, eds., *The Future Control of Food*, 212–220. London: Earthscan.

Taylor, Michael, and Jerry Cayford. 2004. Biotechnology Patents and African Food Security: Aligning America's Patent Policies and International Development Interests. *Minnesota Journal of Law, Science and Technology* 6 (1): 277–304.

Tejera, Valentina. 1999. Tripping over Property Rights: Is it Possible to Reconcile the Convention on Biological Diversity with Article 27 of the TRIPS Agreement? *New England Law Review* 33 (4): 967–988.

Thomas, Gareth. 2005. Innovation, Agricultural Growth and Poverty Reduction. In Calestous Juma, ed., *Going for Growth: Science, Technology and Innovation in Africa*, 74–87. London: The Smith Institute.

UNCTAD-ICTSD. 2003. Intellectual Property Rights: Implications for Development. *Project on IPRs and Sustainable Development.* UNCTAD-ICTSD. http://www.ictsd.org/pubs/ictsd_series/iprs/PP.htm.

United Nations Convention on Biological Diversity (CBD). 1992. Article 8. In-situ Conservation. http://www.cbd.int/convention/articles.shtml?a=cbd-08.

UPOV. 2006. States Party to the International Convention for the Protection of New Varieties of Plants. http://www.upov.int/eng/ratif/pdf/ratifmem.pdf.

van Wijk, Jeroen. 2004. Terminating Piracy or Legitimate Seed Saving? The Use of Copy-Protection Technology in Seeds. *Technology Analysis & Strategic Management* 16 (1): 121–141.

Vivas-Eugui, David. 2003. TRIPS Issues Papers No. 1. Regional and Bilateral Agreements and a TRIPS-Plus World: The Free Trade Area of the Americas Agreement (FTAA). Quaker United Nations Office. http://www.quno.org.

Weiss, Rick. 2008. Firms Seek Patents on "Climate Ready" Altered Crops. *Washington Post*, May 13, A04. http://www.washingtonpost.com/wp-dyn/content/article/2008/05/12/AR2008051209l9_p.

World Intellectual Property Organization (WIPO). 2001. Revised Draft Program and Budget 2002–2003. http://www.wipo.int/documents/en/documents/govbody/budget/2002_03/rev/pdf/pbc4_2.pdf.

World Intellectual Property Organization (WIPO). 2004. Proposal by Argentina and Brazil for the Establishment of a Development Agenda for WIPO. *Document WO/GA/31/11 Presented to the WIPO General Assembly's 31st (15th Extraordinary) Session*. Geneva: WIPO, September 27 to October 5. http://www.wipo.int/documents/en/document/govbody/wo_gb_ga/pdf/wo_ga_31_11.pdf.

World Intellectual Property Organization (WIPO). 2005. Proposal from the United States of America for the Establishment of a Partnership Program in WIPO. *Document IIM/1/2 of the First Session of the Inter-Sessional Intergovernmental Meeting of a Development Agenda for WIPO*. Geneva: WIPO, April 11–12. http://www.wipo.int/search/cs.html?charset=utf-8&url=http%3A//www.wipo.int/edocs/mdocs/mdocs/en/iim_1/iim_1_2.pdf&qt=IIM/1/2&col=meetings&n=2&la=en.

World Trade Organization (WTO), Brazil, India, Pakistan, Peru, Thailand and Venezuela. 2004. Elements of the Obligation to Disclose the Source and Country of Origin of Biological Resources and/or Traditional Knowledge used in an Invention. *WTO Document IP/C/W/429*. September 21.

Wright, Brian, and Phillip Pardey. 2006a. Changing Intellectual Property Regimes: Implications for Developing Country Agriculture. *International Journal of Technology and Globalization* 2 (1/2): 93–114.

Wright, Brian, and Phillip Pardey. 2006b. The Evolving Rights to Intellectual Property Protection in the Agricultural Biosciences. *International Journal of Technology and Globalization* 2 (1/2): 12–29.

8

The Troubled Birth of the "Biotech Century": Global Corporate Power and Its Limits

Robert Falkner

Transnational corporations (TNCs) are powerful drivers of change in the global food system.[1] Nowhere is this more clearly evident than in the field of agricultural biotechnology. The development and commercialization of genetically modified (GM) crops is the fastest technological revolution that has ever occurred in agriculture. In less than two decades, from the 1970s to the 1990s, genetically modified organisms (GMOs) have moved from laboratory research through field testing to commercial production, and since the mid-1990s the global GM planting area has grown at an average rate of 10 percent annually. Led by the US firm Monsanto, a small number of powerful biotechnology firms have set out to reshape global markets for key commodity crops such as soybeans, corn, and canola, with more GM crops (e.g., rice, wheat, potatoes) in the pipeline.

At first sight, agribiotechnology seems to provide ample evidence of overwhelming corporate power in the global food system. But a closer look at the emergence of the GM food business reveals a more nuanced and complex picture. Indeed, the growth of agricultural biotechnology has not been straightforward, and has been met with resistance from consumers, food producers, retailers, farmers, and regulators. In Europe, the majority of consumers rejected GM food when it became available for the first time in 1996, and opposition to agribiotech has also sprung up in Latin America, Africa, and Asia. In the developing world, many countries are still weighing the pros and cons of adopting GM technology in their agricultural systems, and some have even turned down offers of GM food aid despite food shortages. GM food became one of the bêtes noires of the antiglobalization movement of the late 1990s, and continues to ignite heated public debates on how to control the power of TNCs such as Monsanto. In Europe, a system of precautionary GMO regulations has been put in place to carefully test and monitor the environmental and health risks of GMOs, and other nations have also created their own safety regulations and laws. At the international level, a treaty on safety in GMO trade, the Cartagena Protocol on Biosafety, has entered into force, despite opposition from powerful business groups and a coalition of agricultural export countries led by the United States (for an overview of the global politics of GM food, see Falkner 2007a).

Measured against the industry's ambition and rhetoric of transforming global agriculture (see Williams, chapter 6, this volume), it is fair to conclude that "the agricultural biotechnology revolution appears to be a mixed success to date" (Andrée 2005, 135). While it is still early in the history of this new technology, the troubles that have afflicted the GM food revolution raise interesting questions about the power of business in global food governance and in international political economy more generally: Does the ongoing expansion of global GM crop production support the widespread claim that global corporate power is out of control and that social and political actors are failing to direct and shape technological innovation? Or does the worldwide mobilization of antibiotech forces and the creation of stringent biosafety regulations at national and international levels suggest that social and political checks on international business and technology are working? Does the temporary closure of major European and Asian agricultural markets to GM food products demonstrate that global political space exists in which at least some degree of control over corporate power can be exercised?

This chapter seeks to engage with these questions by focusing on the political agency and power of corporations in the international politics of GM food. It examines recent cases of contestation that have slowed down, or at least redirected, the seemingly unstoppable march of biotechnology. This chapter focuses on the role played by tensions and conflict within the corporate sector: between biotech firms on the one hand and food retailers, agricultural traders, and farmers on the other; and within the biotechnology sector itself. By examining business conflict in the evolution of agricultural biotechnology, the chapter aims to contribute to a better understanding of the globalization of the agrifood system and the complex role that corporations play in that process. Business conflict, it is argued, serves to limit the power of the corporate sector and opens up political space for other actors to shape the future of agribiotechnology.

In discussing corporate power in agrifood governance, this chapter builds on the tripartite framing of power as outlined in the introductory chapter by Clapp and Fuchs. It goes beyond this framework, however, by positing the need to focus on the divisions that exist within the business community. Discussions about the relative influence of states, corporations, and societal actors in global governance tend to view corporate power as an aggregate phenomenon, contrasting it with public or social power. Of course, few analysts would naively presume that business is always united in its approach to global governance. But without an explicit acknowledgment that the business community is (potentially) fragmented, and that dynamics of corporate competition and conflict shape the involvement of corporate actors in international politics, the study of global governance would miss out on an important driving force of global change. The aim of this chapter, therefore, is to connect the discussion of business in global governance with the study of business conflict, thereby laying the ground for a neopluralist perspective on corporate power.

The analysis proceeds in four steps. The first section introduces the rise of agricultural biotechnology and gives a brief overview of the emergence and transformation of the biotech industry. The second section sketches the neopluralist perspective that informs the analysis in this chapter, with a special focus on the business-conflict model. The third section introduces three case studies of how business conflict has shaped the path of biotechnological innovation and commercialization. The fourth and final section summarizes the key findings.

The Birth of the "Biotech Century"

Modern biotechnology allows the targeted manipulation of genes in living organisms, which in turn has opened up a vast range of commercial applications in several industrial sectors, from pharmaceuticals to agriculture and environmental remediation. Agriculture emerged as an important focus of biotechnological innovation in the 1980s, as a growing number of small and medium-sized companies were experimenting with different uses of recombinant DNA techniques in plants and animals. Genetic engineering has allowed scientists to insert desirable traits into living organisms or to make plants resistant to pests, drought, herbicides, or other environmental stresses. Several different GM plants have since been developed, most notably soybeans, cotton, corn, canola, and rice.

The global market for GM crops is valued today at over $6 billion (Davoudi 2006), and is second only to medical appliances in modern biotechnology.

The early development of agribiotechnology was led by small and medium-sized companies, many of which had been spun off from research institutes and universities. Over time, with costs of research and commercialization rising rapidly, small biotech firms were gradually taken over by larger life science companies that sought to integrate agricultural and pharmaceutical applications. By the 1990s, the highly fragmented nature of the early biotechnology industry thus gave way to a more consolidated sector with a small number of large industrial players, concentrated in North America, Europe, and Japan.

As GMO innovations were moving from laboratory tests to field trials and commercialization, the process of industrial consolidation took on a new dynamic. From the mid-1990s onward, when the first GM crops were being grown on a commercial scale, a wave of mergers and acquisitions paved the way for a different industrial landscape that saw only a handful of large biotechnology firms dominate GM crop development. The United Kingdom's Astra and Sweden's Zeneca, two large pharmaceutical firms with stakes in agribiotechnology, merged in December 1998 to form the new company AstraZeneca. Only a year later, in December 1999, AstraZeneca and Novartis, the Swiss pharma producer, decided to spin off their respective agrichemical and agribiotechnological businesses and merge them to form Syngenta. And in April 2000, Monsanto and Pharmacia & Upjohn completed a merger of their pharmaceutical operations and created a separate company focused on agribiotechnology, under the name of Monsanto. A key factor behind this wave of mergers was the desire to achieve synergies particularly in the pharmaceutical sector and to broaden the application of genetic engineering techniques to other areas (Fulton and Giannakas 2001).

But the creation of Syngenta and Monsanto with a sole focus on the crops business also suggested that the agricultural and medical sectors were increasingly going their own ways. Against the background of a worsening public climate for GM crops in the late 1990s, combined medical and agricultural biotech firms were keen to separate out the different social, political, and economic risks involved in biotechnology and shield medical applications from the public controversies that began to engulf agricultural firms (King, Wilson, and Naseem 2002).

A different motivation lay behind the second wave of mergers that saw DuPont acquire Pioneer Hi-Breed in 1997, to become the world's largest seed company. Monsanto had kicked off this wave when it decided to take over DeKalb in 1996, and followed this up with further acquisitions in the late 1990s, including Holdens, Delta & Pine Land Co., Asgrow, and Agracetus (Joly and Lemarié 1998). Both DuPont and Monsanto pursued these acquisitions as part of a broader strategy of integrating crop development, agrichemical production, and seed distribution. This, they hoped, would give them greater control over the entire seed and agrichemical business and would put them in a strong commercial position as the sole suppliers to farmers in key markets. The industry is still far away from this vision but has already achieved oligopolistic control over the supply of key GM crop varieties (e.g., soybean, cotton) in countries such as the United States and Argentina (on the latter, see Newell, chapter 9, this volume). Today, after a process of continuous industry consolidation, less than a handful of companies control the global market for GM crops, with Monsanto being by far the dominant player. In 2005, GM crops were grown on an estimated 222 million acres around the world. Monsanto's GM crops accounted for more than 90 percent of the total biotech acreage, followed by the next largest biotech companies, Syngenta, Bayer, and Dow/Du Pont (Davoudi 2006).

Despite a decade of year-on-year growth of GM crop cultivation, the biotech revolution has so far failed to spread worldwide. The global GM crop area has grown steadily for the last ten years, at an average annual rate of around 10 percent. Still, the majority of all commercially grown GM crops can be found in only a handful of countries: the United States, Argentina, Brazil, Canada, and China (James 2006). The United States alone accounts for over half of the world's GM crop production. In contrast, large agricultural markets such as India and China are continuing to debate whether to allow large-scale commercialization of the full range of GM crops that are currently in use. And many of the key import markets, such as the European Union (EU), Japan, and Korea, have put in place stringent import regulations, including GMO labeling requirements and partial or outright bans on GMO imports. Moreover, in many countries where certain GM crops have been authorized for commercial sale, consumers and food retailers are refusing to buy or stock GM food products. The position of importing countries concerned about GMO trade has been further strengthened by the entry into force of the Cartagena Protocol on

Biosafety, an international treaty aimed at ensuring that countries have the right and the capacity to subject GMO imports to risk assessment and to impose precautionary import bans if necessary (Gupta and Falkner 2006).

It would seem, therefore, that the arrival of the "biotech century" (Rifkin 1998) is anything but a straightforward story. Agricultural biotechnology has encountered serious resistance in key markets and has been subjected to increasingly stringent regulations despite an initially favorable regulatory environment in the industrialized world. To understand this crooked path of biotechnological innovation and adoption, we need to consider the role played by different actors—political, societal, and economic—in the shaping of the technology's political-economic environment. Within this field of political contestation, the tensions and conflict between different business actors have played a particularly important role, for they have opened up political space for other actors, most notably environmental nongovernmental organizations (NGOs) and consumer groups, to seek to influence the course of biotechnological commercialization. In the following section, I briefly introduce the notion of "business conflict" as an analytic tool for the study of corporate power, before examining three cases of business conflict in agricultural biotechnology.

Business Power and Business Conflict: A Neopluralist Framework

Economic globalization is generally considered to have strengthened the position of corporations in the international political economy. Most analysts would agree that the combination of greater global economic integration and the spread of liberal market-oriented policies have enhanced the legitimacy of the global corporation and provided it with greater room for maneuver. Assessments of corporate power vary widely, however. For some, corporations "rule the world" (Korten 1995) while state authority is "in retreat" (Strange 1996). Others point to the continuing resilience of the nation-state and international institutions (Hirst and Thompson 1996) and argue that globalization is dependent on a supportive political environment (Waltz 2000).

Part of the problem with debates on business power in an era of globalization is that they have often been conducted at too general and abstract a level. Particularly problematic has been the tendency to force such discussions into the straightjacket of a "zero-sum game," in which rising business power is seen to result in the decline of state power and autonomy. That globalization has changed the environment within which states and firms operate is widely acknowledged. But whether it has resulted in a wholesale transfer of power from public to private actors is less clearly evident. For example, existing trends toward the privatization of global governance (Cutler, Haufler, and Porter 1999) do not necessarily support the broader claim that private power now trumps public power. It can be argued, instead, that private authority is closely linked to, and embedded in, the wider political framework provided by states (Falkner 2003).

Moving beyond the dichotomous positions that have characterized the early debate on globalization would allow us to develop a better understanding of business power—of its nature and sources, as well as its limits. Business may play a more prominent role in an era of globalization, but exactly how business power plays out internationally, and how it affects global governance, is a less straightforward story than either hyperglobalists or globalization skeptics suggest. We need an analytic perspective that views business power in its issue-specific context, as a historically bounded, contingent, concept.

Neopluralism offers this perspective (for a fuller exposition, see Falkner 2008, chapter 2). It acknowledges that business power can take on different dimensions—relational, structural, and discursive—but argues that these need to be contextualized and studied in specific policy domains. Due to their central role in the global economy, business actors may be in a privileged position, vis-à-vis states and nonstate actors, but this alone does not allow them to determine international policy outcomes. The existence of countervailing forces, both in the corporate world and beyond, limits business influence overall. Such forces can be found in the resilience of state power and the proliferation of transnational societal actors, but most importantly in the heterogeneity of the business sector itself. The neopluralist perspective directs our attention to the complex interplay between different actors in global governance, and particularly to the potential for divisions and conflict within the business sector. Neopluralism argues that the diversity of business interests, combined with the persistence of business conflict, constrains business power in the international system and renders global politics open-ended.

That the business community is not always united in its approach to international politics is hardly a novel insight, even though structuralist approaches in the field of international political economy have tended to downplay its significance. Existing approaches that focus on the domestic sources of international politics have long argued that business should not be seen as a monolithic bloc with a pre-given interest. Sectoral differences have been shown to have a profound effect on trade policy in leading industrialized countries (Frieden 1988; Milner 1988), for example, and other divisions among business actors have had a direct or indirect impact on international politics (see the contributions to Cox 1996).

The insights of the business-conflict model are particularly relevant to the study of international environmental politics, and global governance more generally. For business conflict arises whenever international policies or regulations cause changes to existing market structures or create new markets, and thereby cause differential effects on individual business actors. Several lines of conflict can be identified with regard to international regulation, norm setting, and regime building:

- *Between international and national firms* International firms are more likely to favor international rule setting and to support efforts to harmonize different national regulations at the international level than their national counterparts are.
- *Between market leaders and laggards* Technologically advanced firms may derive a competitive benefit from international regulation vis-à-vis other firms that would face higher costs of compliance.
- *Between firms at different points in the production or supply chain* Producers and users of regulated goods may have different interests with regard to the form and content of regulation, and firms operating at the consumer end of the supply chain often face different pressures and demands compared to those more removed from consumer markets.

The specific nature and form of business conflict will depend on the circumstances of the particular issue at hand and the industry or industries concerned. To delve deeper into the potential for business conflict in agribiotechnology, we therefore need to consider in more detail the network of corporate actors that are involved in the research, development, growing, processing, and sale of genetically modified crops and food.

Business-Conflict Formations in Agricultural Biotechnology

A useful starting point in the study of business power and business conflict is to identify the production chain, or supply chain, that links different economic activities in the production, distribution, and sale of a certain good or service. In an age of economic globalization, such production chains are usually transnationally organized, involving producers, traders, and retailers operating in different national locales and across national boundaries. The concept of "production chain" (Dicken 2003, 14) or "commodity chain" (Gereffi and Korzeniewicz 1994), often also referred to as "production network" (Levy, 2008), signifies that different corporate activities are functionally integrated so as to form an international economic network. Such chains or networks can be found in a wide range of economic sectors, from automobiles, aircraft, and computers to textiles and consumer electronics.

Gereffi (1994) distinguishes between two types of networks: *producer-driven networks*, in which multinational corporations centrally control the production system across the entire production chain, as is the case in technology-intensive industries (e.g., Airbus, Siemens); and *buyer-driven networks*, where large retailers and brand-based trading companies play a key role in managing decentralized corporate networks in different national markets (e.g., Wal-Mart, Nike). While this distinction is overly simplistic when compared with the often-complex reality and variety of international economic networks, it helps to identify key structural features of the global economy. Production-chain analysis highlights the international integration of modern industrial production, points to structures of power within the corporate sector, and identifies points of access for social forces that seek to change economic behavior in the global economy.

The global production chain of GM food (see figure 8.1) contains a large number of corporate players from the biotechnology, farming, commodities trade, food production, and retailing sectors. Viewed from the consumer end of the chain, the GM food chain appears to be a buyer-driven production network, where large food retailers exercise considerable influence over the sourcing and production of food products. Such buyer-driven food networks are usually nationally or regionally organized, based on national or regional patterns of retailing and consumption. At the same time, however, we can also view the GM food chain from the producer end, where producers and distributors of GM seeds have undergone a process of industrial concentration. As mentioned earlier, the producer end of the chain has recently seen a trend toward more centralized control by biotechnology firms, with agrichemical and biotechnology companies developing an integrated model of seed production and marketing.

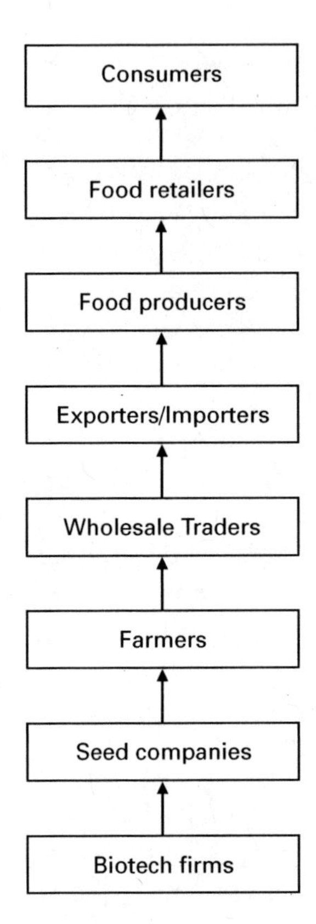

Figure 8.1
The global GM food chain

But efforts to extend corporate control over the GM food chain have largely failed. The highly fragmented farming sector and the diversity of national and regional agricultural markets pose serious barriers to the further integration of the GM food chain from both ends. Neither food retailers nor seed producers have managed to establish full control over the global production and distribution of GM and non-GM food.

What is interesting to note here is that GM food is part of a long production chain that is transnationally organized and that combines different types of corporate actors with diverse interest structures. This in turn offers multiple entry points for social and political actors to seek to influence the production of and trade in GM food products and to form political alliances with corporate interests that are not necessarily aligned with the strategies of the biotechnology industry. As Schurman

(2004) has shown, environmental campaign groups have exploited these so-called industry opportunity structures contained within the GM food chain to press for global change. The following three cases studies seek to demonstrate how such business conflict has shaped the evolution of biotechnological commercialization and have provided major points of contestation over the future of biotechnology in food production.

Three Business-Conflict Formations in Agribiotechnology

Biotech Firms versus Food Retailers

Civil society protests and consumer opposition have severely dented the commercial prospects of biotechnology, particularly in food production. Public unease about genetic engineering surfaced for the first time when GM crops were being developed and field-tested in the 1980s. After the first GM food products became available in commercial quantities in the mid-1990s, antibiotech campaigns erupted in Europe but also in other parts of the world (Bauer, Gaskell, and Durant 2002). Most governments responded by introducing new, or strengthening existing, safety regulations and in some cases even introduced bans or moratoriums on the commercialization of GM crops and food. The United States and Canada were notable exceptions to this trend, and North American consumers generally show much greater acceptance of GM food products. But in many parts of the world, including some of the major import markets in Europe and East Asia (e.g., Japan, Korea), GM food is largely absent from people's daily diet. Rather than transforming global agriculture and food production, as some early biotech pioneers had hoped, GM products have to date conquered only certain segments of the global food market.

While civil society protests and consumer anxieties were a major factor behind the biotech industry's growing troubles, their impact can only be understood if we consider how divisions within the corporate sector have provided the antibiotech movement with important access points to the global product chain. For it was the food retail sector that came to play a critical role in the early demise of GM food in Europe and East Asia, by amplifying consumer concerns and imposing restrictions on GM content in food products. Supermarkets led the worldwide movement toward labeling GM food products, which many governments have since come to mandate through legislation. In Europe, for example, supermarket chains introduced voluntary labels and even eliminated all

GM content from their own-brand food products, even though none of these actions were required by law. Large supermarket chains used their market power over food manufacturers and suppliers to demand the identification, and in some cases elimination, of GM content in food production and distribution, which in turn sent strong signals further down the chain to agricultural traders and farmers. By responding to antibiotech protests and consumer concerns in this way, supermarkets thus raised the hurdles for broader market acceptance of GM food.

Supermarkets have been able to exert this influence because of their increasingly dominant position at the buyer end of the food chain. The overwhelming trend in the European food retail sector has been one of greater concentration, growing market share of own-label products selling under retailer rather than manufacturer names, and increased coordination between retailers and suppliers along the supply chain (Loader and Henson 1998, 32). Leading food retailers have built up a strong position of trust among consumers, particularly with regard to food safety, which enhances their market strength vis-à-vis suppliers but also makes them vulnerable to any form of reputational risk. On issues of critical importance to consumers, therefore, food retailers have tended to emphasize transparency—for example, by providing information on food ingredients through labeling schemes (Loader and Henson 1998, 33).

The short-lived history of GM food in Europe started in 1996, when the first cans of GM tomato puree went on sale. In February, two of the United Kingdom's largest retailers, J Sainsbury and Safeway, offered the clearly labeled GM variety next to non-GM products. Against the background of recent food scares such as over bovine spongiform encephalopathy (BSE, also known as mad cow disease), retailers were keen to ensure that new food technologies would be accompanied by appropriate consumer information (Nunn 2000). But this strategy was under threat from the arrival of new GM crop varieties in the global food chain. The US industry was the first to introduce GM varieties of soybeans and maize to commercial production from 1996 onwards, causing concern among European retailers that these varieties could find their way into the European food chain without the full knowledge of the retail sector. In 1997, European retailers, therefore, called on US commodity suppliers to create segregated distribution channels for GM and non-GM food, so as to prevent accidental commingling. Because this would have caused significant additional costs to US producers and traders, European demands for a dual distribution system were simply ignored. Amidst growing consumer unease and uncertainty regarding a proposed Europeanwide GM food labeling scheme, British food retailers and manufacturers then took the initiative and produced their own labeling code in November 1997. In doing so, the UK retail sector not only preempted legislative steps but also went beyond the measures that were being debated at the EU level. Whereas the EU's draft labeling scheme included references to food products that "may contain" GM content, the UK retail sector's code opted for a more unequivocal form of labeling that indicates that a product "contains" GM content.

Labeling was only a first step, and was unlikely to assure concerned customers. As headlines of "Frankenstein foods" offered a glimpse of what threatened to become a major public relations dilemma for food retailers, some supermarkets decided to distance themselves more clearly from biotechnology products. Iceland, a relatively small retailer with only 1.6 percent of total UK grocery sales, announced in March 1998 that it would eliminate all GM ingredients from its own-label grocery products (Loader and Henson 1998, 33). Sainsbury followed this move in July 1999 and became the first large British supermarket to claim that it had eliminated all GM content (mainly GM soy protein) from its own-brand products. The food retailer worked with more than 1,000 of its suppliers to ensure that only certified non-GM crops would enter the supply chain. Other major British supermarket chains, such as Tesco and Marks & Spencer, were also taking steps to eliminate GM foods from their shelves (Reuters 1999). Under pressure from activist groups, leading supermarkets in Britain, Ireland, France, and Italy then sought to coordinate their efforts and formed a consortium to increase their leverage vis-à-vis GM soy producers (Milmo 1999). They all took voluntary action to eliminate GM content from their food range and put pressure on their suppliers to do the same, or at a minimum to identify all GM food content in the production and distribution chain.

To some extent, European supermarkets acted preemptively, in anticipation of a European directive on GMO labeling (Novel Food Regulation), which had been delayed due to internal disagreements within EU institutions and eventually came into force in 1997 (Loader and Henson 1998, 33). But the main impetus for the introduction of voluntary GM labels, particularly in the United Kingdom, was the fear that rising antibiotech consumer sentiment would dent the retail sector's reputation for food safety. Ironically, opinion polls conducted before GM food labels began to appear in the United Kingdom showed no strong opinion on either side of the GM food debate. According to a Food and Drink Federation opinion survey in 1995, "Only a small proportion [of consumers] indicate that they definitely would/would not buy GMO-containing food products if they were available" (Loader and Henson 1998, 31). Because concern over hostile consumer reaction, however small, was the retail sector's main concern, many large supermarkets went beyond existing regulatory requirements and eliminated all GM content from own-label products. As a consequence, by 2005 twenty-seven of the thirty top European retailers had adopted a policy of excluding GM ingredients from products in their European or main markets (Greenpeace 2005).

Despite being initially open-minded about GM food products, leading European supermarkets ended up hastening the demise of GM food in Europe. The lack of unity between biotechnology firms, agricultural producers, and food retailers has thus dealt a severe blow to industry hopes for a global rollout of GM food. Due to their economic size and exposure to reputational risks at the consumer end of the food chain, food retailers have played a central role in relaying consumer preferences and societal values to food producers, traders, and farmers. In doing so, they may amplify and even distort such preferences and values. In any case, food retailers have a pervasive influence over agricultural practices further down the commodity chain. Crucially for the case of GM food, this influence extends across national boundaries into different farming sectors around the world, causing food producers and traders to seek refuge in strategies of regional market differentiation and segregated distribution. Business conflict has thus led to an increasingly segmented, and ultimately more limited, global market for GM food.

Biotech Firms versus Farmers and Traders

While food retailers transmitted antibiotech consumer sentiment down the GM food chain, it was farmers and agricultural traders that took key decisions on whether to adopt particular GM crop varieties offered by biotech companies. In agricultural markets that are sensitive to the purchasing power of food retailers, those decisions have played an important role in directing the pattern of GM crop cultivation. There are, of course, a variety of reasons why farmers may choose not to adopt GM crops. The higher price of buying a combined package of GM seeds and herbicide treatment may act as a deterrent, and farmers may be doubtful about the long-term economic benefits that can be derived from new and more expensive GM crops. They may also object to the growing dominance of a small number of biotech firms that are increasingly controlling the production and distribution of key seeds. Moreover, they may want to avoid being shut out of export markets with GMO restrictions in place. All these factors have played into farmers' decisions on GM crop cultivation around the world, but it was particularly the latter—the impact of negative consumer reactions and supermarket restrictions—that have come to block the further spread of GM food crops.

A prominent example of how farmer resistance can slow down or temporarily stop the commercialization of GM crops—even in the heartland of the biotech revolution, the United States—is the case of the planned introduction of GM wheat. In 2002, Monsanto submitted applications in Canada and the United States for regulatory approval of Roundup Ready wheat. Only two years later, in 2004, the company was forced to withdraw these applications and publicly declared its intention to stop research and development (R&D) of GM wheat. While this withdrawal may be reversed in the future, the dramatic turnaround in Monsanto's strategy nevertheless demonstrates the impact that consumer preferences, when combined with business conflict, can have on global agriculture. As Olson (2005, 164) put it, "Because of unwavering market rejection of genetically engineered wheat among most international buyers, and because of dogged grassroots opposition from farmers, rural communities and consumers, Monsanto has temporarily corked the genetically engineered wheat genie in the bottle." To understand how this was possible, we need to examine the specific political economy of wheat, and particularly the tensions between the biotechnology industry and wheat farmers and traders in North America.

After its first success with GM soybeans and corn, which captured about 90 and 50 percent market share in the United States respectively, Monsanto shifted its R&D efforts to the next generation of GM crops. In the late 1990s, it was leading the race to develop GM wheat and began to prepare the commercial rollout of a genetically modified version of the high-value hard red spring variety, which is popular with flour mills because of its higher protein content. Monsanto's herbicide-resistant wheat was to be sold in conjunction with its Roundup herbicide, based on the similar agrichemical model of Roundup Ready canola, soybeans, and corn. Hoping to repeat the economic success of earlier GM crops, Monsanto in 2002 sought regulatory approval for the commercial planting of GM wheat in the United States and Canada, and submitted similar requests to regulatory authorities in other countries, such as Japan and South Africa (AP 2004).

By this time, regulatory applications for new GM crops had attracted the attention of environmental campaigners and were starting to meet with resistance from wheat farmers and traders. Farming organizations in Canada and the United States expressed their concern about the negative impact the new crop variety would have on wheat markets. Their objections were based on fears that Monsanto's Roundup Ready wheat might create "super weeds" that would become immune to certain herbicides, and that weed management would become more complicated and costly. They also feared dependence on Monsanto's increasingly monopolistic position in the supply of seeds. Most importantly, wheat farmers with a large exposure to export markets argued that the introduction of GM wheat would cause the loss of important export markets in Europe and Asia because supermarkets there were unlikely to stock GM products.

The growing rift between the biotechnology industry and wheat growers became all too clear when market studies conducted in the early 2000s pointed to persistent resistance to GM wheat in foreign markets. In 2000–2001, eight of the top ten importers of North American hard red spring wheat were based in Asia and Europe, and the overwhelming majority of large buyers in Japan, Korea, China, and the European Union had publicly declared that they would not purchase GM wheat or GMO-contaminated conventional wheat shipments, even if such varieties had passed the regulatory process in their markets (Olson 2005, 157–158). As a representative of Japan's Flour Millers Association, which mills about 90 percent of domestic wheat, stated in 2001: "Under the circumstances, I strongly doubt that any bakery and noodle products made from genetically modified wheat or even conventional wheat that may contain modified wheat will be accepted in the Japanese market. World wheat supply has been abundant in recent years, and I don't see why we have to deal with modified wheat. . . . I believe the production of modified wheat at this time will be a very risky challenge for U.S. producers" (Cropchoice 2001).

Economic studies of the short-term economic impact of GM wheat introduction confirmed the fears of many farmers. Iowa State University economist Wisner warned in his widely noted 2003 study that consumer hostility to GM foods in Europe and elsewhere posed a serious economic risk to American farmers if GM wheat was introduced commercially. Even if production of GM and non-GM varieties were to be segregated, the possibility of accidental commingling and GM contamination would put export sales at risk. With the loss of markets in Europe and Asia, excess American wheat production would quickly lower producer prices and thus make the adoption of GM wheat economically unviable (Wisner 2003).

The political economy of wheat posed a peculiar problem for the biotech industry. Because wheat is grown primarily for food production and human consumption (unlike soybeans and corn, which are predominantly processed into animal feed or intermediate products such as oils), it is much more susceptible to antibiotech consumer attitudes. Also, American producers grow a much smaller share of global wheat production (8 percent in 2002) and therefore face much stiffer international competition in export markets. If US wheat production were to switch to GM varieties, US wheat exports could easily be replaced with non-GM supplies from other producer countries. And finally, US wheat farmers had already experienced a long-term downward trend in their global market share, down from nearly 50 percent in the 1970s to just over 20 percent in 2001 (Wisner 2003). All these factors combined to heighten the threat of losing important export markets.

Sensing growing resistance among North American farmers, Monsanto set up a Wheat Industry Joint Biotech Committee in 2001 to seek a consensus with the farming community on how to introduce its GM wheat variety. At the same time, it set the period of 2003–2005 as the window for commercial introduction of GM wheat while assuring farmers that such a decision would depend on market acceptance for the biotech crop (Fairchild 2002). But concerned wheat producers were actively seeking moratoriums and even bans on GM wheat commercialization in those states with the highest stakes in wheat production, such as North Dakota and Montana.

To be sure, not all wheat farmers were opposed to GM wheat. Those producing primarily for the domestic market have remained open-minded about the potential commercial benefits of biotechnology. The wheat farming sector was not united on this question, with the National Association of Wheat Growers (NAWG) being more sympathetic to GM crops and the US Wheat Associates (USWA) voicing the concerns of exporters (Bernick and Wenzel 2006). But in the end, the divisions within the farming community and strong lobbying by exporter interests, who account for over half of the American wheat market (Wisner 2004 17), made it impossible for Monsanto to fulfill its pledge to create a broad market consensus before commercializing GM wheat.

Opposition to GM wheat commercialization ran particularly high among Canadian farmers, who are more dependent on export markets than their US counterparts. The Canadian Wheat Board actively lobbied regulatory authorities not to allow the planting of Monsanto's new crop and in May 2003 appealed directly to the company to withdraw its application for regulatory approval. The Canadian farmers' stance was based almost entirely on an economic rationale. Because of the danger of commingling of GM and non-GM crops and the high costs of segregated production and distribution systems, wheat growers feared the loss of access to high-premium markets abroad, greater uncertainty in international farm trade, and higher costs for farm management and grain handling (Johnston 2003).

By 2004, it had become clear to Monsanto's executives that their strategy of building a consensus among North American farmers had failed. Not wanting to further antagonize the farming community, the company announced in May that it was delaying plans for the commercial introduction of Roundup Ready wheat and that it would redirect its R&D efforts toward other crops such as corn, oilseeds, and cotton. Executive Vice President Carl Casale declared that "as a result of our portfolio review and dialogue with wheat industry leaders, we recognize the business opportunities with Roundup Ready spring wheat are less attractive relative to Monsanto's other commercial priorities" (Burchett 2004). Monsanto has since worked hard to persuade North American farming associations to support in principle the commercial introduction of GM wheat at some future point (Bernick and Wenzel 2006), but has had little success, particularly with Canadian farming associations.

Monsanto's failure to introduce GM wheat in North America, the most receptive market environment for GM crops, is a prime example of how business conflict along the product chain has offered antibiotech forces important access points to shape the future of biotechnology. Such "industry opportunity structures" (Schurman 2004) emerged in a number of areas, wherever GM commercialization threatened to increase the costs of crop management, storage, and distribution, and particularly where it would cause the loss of export markets with GMO import restrictions or labeling requirements in place. In China, for example, exporters of soybeans and soy-based products to Japan, Korea, and the

EU have lobbied the government against the start of commercial production of GM soybean varieties in the country, out of fear that commingling and accidental GM contamination of shipments abroad would close off foreign markets (Falkner 2006). In similar fashion, Indian rice exporters have resisted plans to test and introduce GM rice to Indian agriculture. In November 2006, the All-India Rice Exporters' Association called for an immediate end to all field trials of a GM rice variety developed by Monsanto-Mahyco, for fear of losing its position as the world's third largest exporter of basmati rice (Parsai 2006). Consumer resistance and environmental protest are of course the root cause of the industry's difficulties, but in all such cases it is business conflict that enabled antibiotech movements to shape the commercial future of emerging GM crops.

Business Conflict within the Biotech Sector

So far, the focus has been on tensions between biotech firms and other corporate players in the global GM food chain. This is in line with traditional business-conflict studies in international political economy, which emphasize cleavages between different industrial sectors, particularly between domestic and more internationally oriented sectors (e.g., Frieden 1988; Stant 1996). In international environmental politics, such interindustry conflicts are common where regulations have distributional effects between industries that cause pollution and those that provide environmental solutions (e.g., fossil fuel versus renewable energy industries in climate change; see Falkner 2008, chapter 4). But business conflict can also emerge within an industry—for example, between technological leaders and laggards, or between dominant players and new entrants to a given market. Can similar intraindustry divisions be found in the biotechnology sector?

At first glance, the answer seems to be no. All biotech firms share a common interest in promoting the use of biotechnology in agriculture and in removing regulatory barriers to technological innovation. They may be locked into fierce economic competition when it comes to developing new products, capturing market share, or seeking regulatory approval for their own products, but their basic political outlook on larger questions of biotechnology governance can be expected to be similar. Indeed, a decade of commercial GMO production and nearly two decades of international biosafety politics suggest that the divisions within the biotech sector have been less pronounced than those between the biotechnology sector and other sectors involved in the GM food product chain.

Yet, certain fault lines within the biotech sector did emerge in the context of the international negotiations on the Cartagena Protocol on Biosafety. One potential source for business conflict could be found in the different outlooks and lobbying strategies of US versus European biotech firms, though such differences never seriously threatened their common lobbying position. The other, more significant, cleavage existed between the different economic and political strategies of agricultural versus medical biotech firms, which came to undermine the structural power of the biotech sector overall. None of these cleavages developed into a full-scale business-conflict constellation, although the growing disengagement between the agricultural and medical sectors was to have a lasting effect on the strength of the business lobby in the international negotiations.

Let us consider regional difference first. US and European companies developed distinctive environmental strategies in the past, as has been noted in the international politics of climate change and ozone layer depletion (Levy and Newell 2000). How important were such transatlantic differences in the biosafety negotiations? Participants in the Cartagena Protocol talks have attested to some regional variation in the lobbying role of biotech firms. While US corporate representatives were more confrontational in defending their antiregulatory stance, European lobbyists tended to strike a more conciliatory note and were more willing to consider some form of international regulation.[2] This impression was shared by environmental NGOs who participated in the biosafety negotiations. Richard Tapper (2002, 271), an advisor to the Worldwide Fund for Nature, remarked that European biotechnology representatives "were more open to constructive dialogue than some of their US counterparts." European negotiators likewise commented that they encountered a more hostile lobbying stance by US industry representatives than by their European counterparts. This was partly a question of lobbying style and cultural differences, but also reflected different regulatory pressures and relations with home governments.

But did these regional differences in lobbying translate into a significant business conflict? Were US and European biotech firms divided by fundamentally different political strategies? There is little evidence to support such claims (see Clapp 2007). The main reason for this is the high degree of internationalization in biotechnology that has helped to blur regional differences between individual companies, in terms of both global commercial strategies and industry structures. All biotech firms operate in a global market. They have developed similar biotech products, have sought regulatory approval for their GM crops in the same markets, and have faced the same political constraints. While they have needed to adjust their strategies to local and regional conditions, they have nevertheless maintained the same overall strategic outlook. Levy and Newell (2000, 14) are therefore right to argue that, "despite procedural and institutional influences on the way businesses have pursued their interests and different degrees of exposure to social and political concerns about GM foods, biotech companies in both Europe and the United States have maintained similar positions."

The global biotech industry structure, which is characterized by growing internationalization of research and development, has also rendered regional differences less relevant. The series of mergers and acquisitions described above has led to a dramatic concentration of agribiotechnology in the hands of only a few European and North American firms. Even though these firms retain strong links to their domestic markets in the United States, Germany, and Switzerland, in fact R&D, field testing, and commercialization of new crops are highly internationalized. European biotech firms such as Syngenta and Bayer have a strong presence in the North American market and are in close competition with Monsanto in overseas markets. Because of this high degree of industrial internationalization, it makes only limited sense to speak of Bayer and Syngenta as representing "European biotechnology" when it comes to international biosafety politics. Indeed, as a US industry lobbyist remarked, "There was no European industry position as such" in the negotiations on the Cartagena Protocol.[3]

Regional differences may have played only a marginal role, but sectoral differences within the biotech industry had a more profound effect on the international process. During the 1990s, the reorganization of the biotechnology industry led to a growing separation of firms focused on either agricultural or medical applications of genetic engineering, which in turn led to the emergence of divergent political strategies. One of the key contributing factors to this growing divergence was the varying levels of public resistance that different forms of biotechnology faced. As antibiotech sentiment against GM food—but not GM pharmaceuticals—came to pose a growing commercial threat, companies that straddled both these sides of biotechnology began to disentangle their respective businesses so as to separate out the different political risks attached to food and medicine. This process took off in the late 1990s and culminated in two large mergers that transformed the industrial landscape: Novartis' and AstraZeneca's decision in 1999 to combine their agricultural business in the newly formed but separate company Syngenta; and the merger of Monsanto and Pharmacia & Upjohn in 2000, which led to the creation of a separate agribiotech company under Monsanto's name.

The growing separation of agricultural and medical biotechnology was clearly visible in the biosafety negotiations. Pharmaceutical companies had only a limited interest in the international process and lobbied mainly to have GMOs for pharmaceutical purposes excluded from the biosafety protocol (Marquard 2002). Rather than working together with agribiotech firms and forming a united industry front, the pharmaceutical industry focused solely on limiting the scope of the protocol to exclude the medical sector. Representation by pharmaceutical firms in the negotiation was limited in any case, and many of their representatives departed early from the final negotiation round in January 2000, once it was decided that pharmaceutical GMOs would not be covered by the protocol. The disintegration of agricultural and medical biotechnology in the late 1990s thus severely undercut the overall lobbying effort of the industry. It made it easier for the European Union to champion precautionary biosafety regulations on agricultural GMOs despite its continued support for biotechnology in the medical area (see Falkner 2007b).

In this way, business conflict among biotech firms came to play a significant role in the international political process. While regional differences were less important to the overall lobbying position of the industry, the growing separation of agricultural and medical biotechnology severely restricted the industry's overall influence. With its larger pharmaceutical allies departed from the international negotiations, the agribiotech industry increasingly had to rely on links with agricultural trading firms to fight burdensome regulations in the emerging biosafety regime.

Conclusion

The much debated arrival of the biotech century has begun to transform major industrial sectors, including agriculture, where it provides new avenues for crop development and integrated crop management. But despite the hopes of early industrial innovators, the adoption of modern biotechnology has been much slower in agriculture than in the medical sector. Genetically modified crops have run up against widespread resistance, by consumers, health and environmental campaigners, regulators, and farmers. A few GM crops (soy, corn, canola) have been introduced in a small but growing number of countries (especially the United States, Canada, Brazil and Argentina). Especially in Europe, Asia, and Africa, however, consumer hostility and environmental protests have dented the biotech industry's global ambitions. The worldwide GM crop area may be expanding and new GM crops are being developed, but key food markets remain largely closed to these developments. Of course, it is still early in the history of modern biotechnology, but the uneven adoption of GM crops around the world holds important lessons for the study of corporate power in global food governance.

As the analysis in this and in other chapters has shown, social protest and consumer preferences are potent forces that shape global agricultural markets. Antibiotech sentiment has been on the rise since the mid-1990s and has led to the closure of major markets to GM food products. But to understand why social protest has had such a powerful influence over commercial agribiotechnology, we need to focus on the nature of the global GM food chain and the sources of friction and conflict within it. Because business has been divided on how to assess the economic benefits and risks associated with GM crops and whether to label GM content in food products, antibiotech forces have been able to exploit these divisions and create barriers to the further commercial introduction of GM food. The biotechnology industry has found it difficult to persuade certain farming sectors and food retail businesses of the benefits of its novel products. Despite their success in vertically integrating crop development and seed distribution, leading companies such as Monsanto have been unable to penetrate major agricultural markets. In short, business conflict has frustrated the global ambitions of the biotech industry, despite its seemingly unstoppable push for a biotech revolution in world agriculture.

That latent divisions between different business actors have developed into full-fledged business conflict is largely due to the rise of antibiotech sentiment and protests. Societal and political actors have actively sought to exploit the potential for business conflict and have amplified divisions within the business sector. Business conflict has opened up opportunities for more effective social and political contestation and has provided antibiotech forces with levers of influence over the direction of GM crop commercialization, thus creating a more open-ended and indeterminate path of biotechnological development. In this sense, business conflict limits business power overall, whether structural, instrumental, or discursive. Whether such contestation enhances the possibilities of global food governance for sustainability depends on other factors, however. Business conflict merely creates political space for society to shape technological innovation and commercial developments in agriculture.

Notes

1. An earlier version of this chapter was presented at the 2007 Annual Convention of the International Studies Association in Chicago. This chapter builds on and extends the analysis in chapter 5 of Falkner 2008.
2. Interview with US biotechnology representative, July 16, 2001.
3. Interview with US biotechnology representative, July 16, 2001.

References

Andrée, Peter. 2005. The Genetic Engineering Revolution in Agriculture and Food: Strategies of the "Biotech Bloc." In David Levy and Peter J. Newell, eds., *The Business of Global Environmental Governance*, 135–166. Cambridge, MA: MIT Press.

AP. 2004. Monsanto Plans for Biotech Wheat On Hold. *Associated Press*, May 10.

Bauer, Martin W., George Gaskell, and John Durant, eds. 2002. *Biotechnology: The Making of a Global Controversy*. Cambridge: Cambridge University Press.

Bernick, Jeanne and Wayne Wenzel. 2006. Finally, a Biotech Statement from the Wheat Organizations. *Farm Journal*, March 8.

Burchett, Andrew. 2004. Monsanto Mothballs Roundup Ready Wheat. AGWEB, May 10. http://www.agweb.com/get_article.asp?pageid=108307.

Clapp, Jennifer. 2007. Transnational Corporate Interests in International Biosafety Negotiations. In Robert Falkner, ed., *The International Politics of Genetically Modified Food: Diplomacy, Trade and Law*, 34–47. Basingstoke: Palgrave Macmillan.

Cox, Ronald W., ed. 1996. *Business and the State in International Relations*. Boulder, CO: Westview Press.

Cropchoice. 2001. Global Opposition Mounts against Monsanto's "Franken-Wheat." February 1. http://www.cropchoice.com.

Cutler, A. Claire, Virgina Haufler, and Tony Porter, eds. 1999. *Private Authority and International Affairs*. Albany: State University of New York Press.

Davoudi, Salamander. 2006. Monsanto Strengthens Its Grip on GM Market. *The Financial Times*, November 16.

Dicken, Peter. 2003. *Global Shift: Reshaping the Global Economic Map in the 21st Century*. London: Sage.

Fairchild, Barbara. 2002. Release Date for GM Wheat Is 2005. *Farm Journal*, April 9.

Falkner, Robert. 2003. Private Environmental Governance and International Relations: Exploring the Links. *Global Environmental Politics* 3 (2): 72–87.

Falkner, Robert. 2006. International Sources of Environmental Policy Change in China: The Case of Genetically Modified Food. *The Pacific Review* 19 (4): 473–494.

Falkner, Robert, ed. 2007a. *The International Politics of Genetically Modified Food: Diplomacy, Trade and Law*. Basingstoke: Palgrave Macmillan.

Falkner, Robert. 2007b. The Political Economy of "Normative Power" Europe: EU Environmental Leadership in International Biotechnology Regulation. *Journal of European Public Policy* 14 (4): 507–526.

Falkner, Robert. 2008. *Business Power and Conflict in International Environmental Politics*. Basingstoke: Palgrave Macmillan.

Frieden, Jeffry A. 1988. Sectoral Conflict and U.S. Foreign Economic Policy, 1914–1940. *International Organization* 42 (1): 59–90.

Fulton, Murray, and Konstantinos Giannakas. 2001. Agricultural Biotechnology and Industry Structure. *AgBioForum* 4 (2): 137–151.

Gereffi, Gary. 1994. The Organization of Buyer-Driven Global Commodity Chains: How U.S. Retailers Shape Overseas Production Networks. In Gary Gereffi and Miguel Korzeniewicz, eds., *Commodity Chains and Global Capitalism*, 95–122. Westport: Praeger.

Gereffi, Gary, and Miguel Korzeniewicz, eds. 1994. *Commodity Chains and Global Capitalism*. Westport: Praeger.

Greenpeace. 2005. *No Market for GM Labelled Food in Europe*. Greenpeace International, January. http://www.greenpeace.org/raw/content/eu-unit/press-centre/reports/no-market-for-gm-labelled-food.pdf.

Gupta, Aarti, and Robert Falkner. 2006. Implementing the Cartagena Protocol: Comparing Mexico, China and South Africa. *Global Environmental Politics* 6 (4): 23–55.

Hirst, Paul, and Grahame Thompson. 1996. *Globalization in Question: The International Economy and the Possibilities of Governance*. Cambridge: Polity Press.

James, Clive. 2006. Global Status of Commercialized Biotech/GM Crops: 2006. *ISAAA Briefs No. 35*. Ithaca: ISAAA.

Johnston, Julianne. 2003. CWB Asks Monsanto to Put the Brakes on Roundup Ready Wheat. AGWEB, May 27. http://www.agweb.com/get_article.asp?pageid=98129.

Joly, Pierre-Benoit, and Stéphane Lemarié. 1998. Industry Consolidation, Public Attitude, and the Future of Plant Biotechnology in Europe. *AgBioForum* 1 (2): 85–90.

King, John L., Norbert L. W. Wilson, and Anwar Naseem. 2002. A Tale of Two Mergers: What We Can Learn from Agricultural Biotechnology Event Studies. *AgBioForum* 5 (1): 14–19.

Korten, David C. 1995. *When Corporations Rule the World.* London: Earthscan.

Levy, David L. 2008. Political Contestation in Global Production Networks. *Academy of Management Review* 33 (4)

Levy, David L., and Peter J. Newell. 2000. Oceans Apart? Business Responses to the Environment in Europe and North America. *Environment* 42 (9): 8–20.

Loader, Rupert, and Spencer Henson. 1998. A View of GMOs from the UK. *AgBioForum* 1 (1): 31–34.

Marquard, Helen. 2002. Scope. In Christoph Bail, Robert Falkner, and Helen Marquard, eds., *The Cartagena Protocol on Biosafety: Reconciling Trade in Biotechnology with Environment and Development?*, 289–298. London: RIIA/Earthscan.

Milmo, Cahal. 1999. Sainsbury's Bans GM Food from Own Brand Range. *PA News*, March 17. http://archives.foodsafetynetwork.ca/agnet/1999/3-1999/ag-03-17-99-01.txt.

Milner, Helen V. 1988. *Resisting Protectionism: Global Industries and the Politics of International Trade.* Princeton, NJ: Princeton University Press.

Nunn, Janet. 2000. What Lies Behind the GM Label on UK Foods. *AgBioForum* 3 (4): 250–254.

Olson, R. Dennis. 2005. Hard Red Spring Wheat at a Genetic Crossroad: Rural Prosperity or Corporate Hegemony? In Daniel L. Kleinman, Abby J. Kinchy, and Jo Handelsman, eds., *Controversies in Science and Technology. Vol. 1: From Maize to Menopause*, 150–168. Madison: University of Wisconsin Press.

Parsai, Gargi. 2006. Exporters Seek Ban on Field Trials of GM Rice. *The Hindu*, November 2.

Reuters. 1999. Sainsbury Says Own-Brand Ingredients GM-Free. *Reuters*, July 19. http://archives.foodsafetynetwork.ca/agnet/1999/7-1999/ag-07-19-99-01.txt.

Rifkin, Jeremy. 1998. *The Biotech Century: Harnessing the Gene and Remaking the World.* New York: Jeremy P. Tarcher/Putnam.

Schurman, Rachel. 2004. Fighting "Frankenfoods": Industry Opportunity Structures and the Efficacy of the Anti-Biotech Movement in Western Europe. *Social Problems* 51 (2): 243–268.

Stant, William N. 1996. Business Conflict and U.S. Trade Policy: The Case of the Machine Tool Industry. In Ronald W. Cox, ed., *Business and the State in International Relations*, 79–108. Boulder, CO: Westview Press.

Strange, Susan. 1996. *The Retreat of the State: The Diffusion of Power in the World Economy*. Cambridge: Cambridge University Press.

Tapper, Richard. 2002. Environment Business & Development. In Christoph Bail, Robert Falkner, and Helen Marquard, eds., *The Cartagena Protocol on Biosafety: Reconciling Trade in Biotechnology with Environment and Development?*, 268–272. London: RIIA/Earthscan.

Waltz, Kenneth N. 2000. Globalization and American Power. *The National Interest* (59): 46–56.

Wisner, Robert N. 2003. *Market Risks of Genetically Modified Wheat: The Potential Short-Term Impacts of GMO Spring Wheat Introduction on U.S. Wheat Export Markets and Prices*. Ames: Iowa State University, October 30. http://www.worc.org/pdfs/wisnerfinal.pdf.

Wisner, Robert N. 2004. *Round-Up Ready® Spring Wheat: Its Potential Short-Term Impacts on U.S. Wheat Export Markets and Prices*. Economic Staff Report. Ames: Department of Economics, Iowa State University, July 1.

9

Technology, Food, Power: Governing GMOs in Argentina

Peter Newell

The global governance of food and agriculture takes place in and across many sites within the global political economy. It derives from the quotidian practices of food politics in public, private, and hybrid arenas between a range of social and political actors with competing notions of *how* agricultural production should be organized and governed and *for whom*. Particular sites in the global economy manifest these politics and at the same time construct them. Because of its status as a leading exporter of genetically modified (GM) foods, as a key ally of other biotech superpowers like the United States and Canada, and as the location of a series of conflicts around the ownership and developmental benefits of crop biotechnology, what happens in Argentina has global repercussions, just as global politics sharply define the governance of genetically modified organisms (GMOs) in Argentina.

The story of agricultural biotechnology in Argentina brings to the fore a potent combination of the politics of poverty, the power of multinational corporations, and the political economy of food and agriculture. Because the country is a leading exporter of GMOs, Argentina's politics of biotechnology provide unique insight into conflicts over agricultural futures, contestations over corporate power, and the politics of development. The focus of this chapter is the role of agrifood corporations in the political economy of biotechnology in Argentina. Argentina presents a fascinating case study of the political economy of governing biotechnology for a number of reasons. First, globally, with allies such as the United States and Canada, Argentina is a member of global pro-biotech coalitions. With growing trade ties to China, it continues to be the second most significant cultivator of GM crops in the world. As a proactive developing-country state in the biosafety negotiations and as a proponent of the World Trade Organization (WTO) case brought against the

European Union (EU), Argentina's domestic politics play out globally, just as global politics serve to define the boundaries of autonomy at the national level.

Second, unlike many developing countries whose view of the technology has been informed, at least rhetorically, by concerns with food security, Argentina has embraced the technology on grounds of its export potential. This makes it an interesting case of biotechnology's potential to advance or hinder development.

Third, within Latin America, the country's regulatory system is seen by many as a model to be emulated, such that what happens in Argentina could be "exported" to other countries within the region (Garcia-Johnson 2000).

Fourth, Argentina is seen by investors as a gateway to gain access to the rest of the continent for biotech products. Given this, the politics of Mercosur and the flow of nonapproved GM seeds to Brazil through neighboring Paraguay provide another interesting regional political angle.

Such an analysis makes important contributions to a number of debates addressed in this book. An analysis of state-business relations in one of the world's leading actors in biotechnology speaks to debates about biotechnology policy and politics, to the role of business in particular (Newell 2006, 2007; Glover and Newell 2004; Glover 2007; Falkner, chapter 8, this volume), and to discussions about biotechnology and development and the compatibility of the two. In a context of hunger and a food crisis produced by economic collapse, the debate about biotechnology's role in alleviating poverty has taken on an unprecedented urgency in Argentina.

Moreover, by looking at the different forms of power exercised by firms in the governance of food and agriculture, a contribution is made to emerging theories of the firm in general (Amoore 2000), and to conceptualizations of the role of the firm in global environmental politics in particular (Levy and Newell 2005; Falkner 2007). Looking at how global firms, such as Monsanto, operate in different jurisdictions and policy environments as a way of understanding corporations as political actors in the governance of food and agriculture also deepens our understanding of comparative business studies. This case study helps us understand the link between the economic functions firms perform within value chains and as market actors, and the ways these roles shape their political agendas pursued within national arenas and internationally as key actors in the "public" global governance of food and agriculture. It enables us to more effectively integrate market and technological dimensions of power with political strategy, since ultimately all strategy is political (Levy and Newell 2002).

I argue not only that the role of corporations in the production of food makes them central governance actors in and of themselves, irrespective of the extent to which and ways in which they affect the power of states and international institutions, but that a key relationship exists between production, power, and governance (Newell 2008c). The nature of that relationship in a globalized and highly unequal political economy means that the scope for autonomous and democratic decision making about such critical issues as the production of food, the creation of rural employment, and the sustainability of the prevailing (industrial) organization of agriculture, is bounded and constrained by the structure it seeks to govern. This is not just about the global organization of famines and feasts that Saurin (1997) draws our attention to; the way the power structures of the world economy determine the production and supply of food as well as access to it; who eats, who starves, and why. Rather, it is about how relations of power configure to make certain ways of organizing food production, and the use of particular technologies of production, appear as normal, neutral, and benign when in fact, as with any system of production, their benefits are contestable, their impacts often uneven, and the choices deeply political in terms of their distributional consequences and degree of sustainability.

Explaining the political process of alliance and coalition building and the strategies that are invoked to sustain an environment supportive of a particular way of doing things, demands that we explain not just the *types of power* at work (structural, instrumental, and discursive) as clearly laid out in chapter 1, but that we explain where that power to shape outcomes comes from and why it is some actors have a privileged role in discussions about how we produce and what we eat. I argue that a combination of three interrelated sources of power—material, institutional, and discursive—help to explain why, in the case study explored in this chapter, corporate power takes the form it does in Argentina, whereas in other settings it has been more contested (Newell 2008b).

I argue that differences of interest inevitably exist between businesses depending on their place in the supply chain, where in the process of production they extract the highest value, and, as a result, the types of regulation that affect their operations (Falkner, chapter 8, this volume).

This is what produces the distinct political formations and forms of business representation that I describe in this chapter. It is certainly also the case that the conflicts between different fractions of capital in their attempt to present their interests as those of capital-in-general provide opportunities for the formation of new alliances with elements within civil society or sympathetic parts of the state, as has been documented elsewhere (Levy and Newell 2005). Such an argument does not, however, invalidate the argument that distinct fractions of capital do not, at the same time, have certain interests in common, not least the smooth functioning of a market economy in a way that continues to create new possibilities for capital accumulation in an environment where property rights are allocated and protected (even if the process by which this is achieved is contested), supportive regulation provided, and social unrest contained. Contests over how best to secure the most conducive conditions for capital accumulation among firms underpinned by shared interests in the overall outcome of bargaining should not necessarily be taken as evidence that differences between firms amount to profound political divisions that provide proof of a pluralist politics at work. What sets (neo) pluralist accounts (Falkner 2007) apart from the account provided here is also the theory of the state. Far from being a neutral actor in the governance of food and agriculture, the state is intertwined with the private actors whose interests we are expecting it to regulate, a situation that produces contradictions when public and private power are not so easily disentangled.

This should inform our view about how much policy space we currently have as societies to demand changes and outcomes that may challenge the prevailing structures of production on which political power, in brief, rests. Victories in skirmishes between business and civil society have produced an interesting set of outcomes, often interpreted as an example of the limits of business power (see Falkner, chapter 8, this volume). In the war of position over the battle for agricultural biotechnology (Andrée 2005), leading biotech firms have certainly had to revisit their marketing and corporate strategies and some elements of finance capital looking for quick and easy returns have lost interest, but the global struggle to secure a future for biotech continues. It continues not just in the public multilateral arenas of biosafety negotiations where many conventional approaches would focus their attention, but via the politics of aid as we saw in Zambia, Bolivia, and elsewhere where GM food aid was sent in times of national emergency (see Clapp, chapter 5, this volume). It is conducted through trade forums such as the WTO as the case described below makes clear, or through synchronized measures in bilateral trade agreements such as that concluded between the United States and Peru (Newell 2008b), as well as through struggles by corporations to ally their vision for biotechnology with the interventions of development institutions such as the World Bank and Food and Agriculture Organization. With greater power and resources than their opponents, as Sell shows in chapter 7 of this book, large firms are able to ensure that questions about food and agriculture that affect them are settled in forums in which they have an advantage. In global terms, despite ongoing consumer resistance in some quarters, the year-on-year growth in areas of the world cultivating GM crops suggests that biotech continues on an upward trajectory. Though continued growth is far from inevitable, it does provide pause for thought about how far biotech is a case of the limits of business power.

The case study in this chapter describes one specific site in this broader global terrain over the future of food and agriculture. Following a brief overview of key developments in Argentina, the chapter proceeds with an analysis of the key private actors in this sector—their corporate and policy strategies and forms of association and mobilization. An examination of the sources of corporate power in the governance of biotechnology in Argentina is then developed, focusing on its material, institutional, and discursive dimensions before closing with some tentative conclusions about the implications of these sources of corporate power for alternative ways of organizing the governance of food and agriculture in a context in which state economic strategy is so entirely dependent on private agricultural interests. The research that underpins the chapter combines over twenty-five key informant interviews[1] to date and the use of a wide range of published and unpublished academic resources, gray (non-academic) literatures, and media sources.

Overview: Agricultural Biotechnology in Argentina

Argentina is now the world's second largest producer and exporter of GM crops, accounting for 23 percent of global production. Seven GM crops have been approved for commercialization, including glyphosate-resistant soybean and herbicide and insect-resistant varieties of maize and cotton, all of them in response to evaluations requested by multinational companies. GM soy is Argentina's most extensive GM crop, comprising almost 90 percent of the twelve million hectares planted in 2001–2002 and nearly half of Argentina's total agricultural production by 2002–2003. With a 7 percent increase in the area under cultivation of GM soy for the latest season for which figures are available (2005–2006), the trend looks set to continue amid rising international prices for the commodity. Underscoring the export-driven nature of this model, in 2003, 98 percent of soy was exported in Argentina as beans, soy meal for animal feed, and soy oil, representing about 20 percent of Argentina's total exports by 2004 (Galli 2005). More importantly, it has played a key role after Argentina defaulted on its US$140 billion national debt in December 2001, and an enormous devaluation took place. As Argentine agriculture trader Alejandro G. Elsztain commented: "The IMF [International Monetary Fund] should be very happy with us. Without agribusiness and oil, Argentina would never meet the surplus they are demanding"(quoted in Vara 2005, 8).

The government decided that income derived from exports would help to increase foreign earnings that would in turn help the poor. Revenues earned from taxes imposed on exports of GM soy have been used to subsidize the internal market for food, for example. This issue flared up in early 2008 with proposals to increase the level of taxation on exports amid sharp rises in the price of soy. The move by the new government of Cristina Kirchner has led to big demonstrations by large farming groups and their supporters, in conflict with the movement of the unemployed (*los piqueteros*) who defend the government's claim that wealthy landowners can afford to pay more to support the welfare of the poor. Other claimed benefits associated with the GM crops have been the savings from reduced pesticide use and reduced soil erosion from less intensive tilling. While there is some evidence of these effects, more critical accounts suggest that the record to date has been less positive if a wider range of issues are taken into account, such as evidence of increased chemical imports (such as glyphosate from China), of deforestation associated with land clearing for GM production (Greenpeace 2006),[2] as well as concentration of land tenure and decreasing employment among laborers lower down the agricultural supply chain.

Nonetheless, the fact that nearly all of Argentina's GM production is for export, principally as animal feed, means that, to date, it has avoided much of the public controversy surrounding human consumption of GMOs that has characterized debates in Europe and parts of Asia in particular. Notwithstanding some debate about the pros and cons of a monocrop strategy, there has been little debate on environmental impacts, with the focus much more squarely on the merits of biotechnology as an economic and developmental strategy. Indeed, the Office of Biotechnology of the Secretary of Agriculture—set up in 2004 to coordinate overall policy on agricultural biotechnology—developed a ten-year (2005–2015) *Strategic Plan for Agricultural Biotechnology*,[3] which reaffirmed a critical role for biotechnology as the main source of technological solutions for agricultural productivity growth in the country. The plan proposes to create a favorable environment (in political, legal, and public terms) for the creation and development of biotechnology-based companies, and also for the consolidation of existing companies.

The Organization and Mobilization of Industry

To understand why some corporations or some business coalitions exercise more power than others in the governance of food and technology and why they advance some policy positions over others, it is necessary to identify how businesses organize themselves, which agendas are adopted, which ignored, and to understand the intercapital conflicts that define their relations with one another. This is useful in challenging a view of business as a monolithic actor without losing sight of the common interests that unite corporate actors, which they acknowledge and act on.

There is not sufficient space here to review in any detail all the producer groups concerned with different aspect of agricultural biotechnology supply chains. I focus here on those with most direct interest in transgenics and the governance thereof. This includes ASA (Asociación de Semilleros Argentinos),[4] ArgenBio, Foro Argentino de Biotecnología (FAB),[5] and APRESID (Asociación de Productores en Siembra Directa).[6] Each of these bodies represents businesses in different parts of the biotech production chain (see Falkner, chapter 8, this volume). While ASA serves the needs of larger-scale seed growers, a different fraction of capital again is served by APRESID and ArgenBio which serve as publicity and promotional organizations for the needs of multinational biotech capital with members such as Monsanto, Dow Agrosciences, and Sygnenta but with less presence in official biotech decision-making forums.

ArgenBIO plays an overt public-political role in promoting and defending biotechnology in Argentina on behalf of the key transnational corporations (TNCs) active in this area that make up its key organizational committee. It is linked with the BIO (Biotechnology Industry Organization) offices that exist throughout the world for the same purpose and receives its funding from Crop Life International.[7] It is more globalized than other industry groups, given its source of funding and who it represents, which has meant its representatives have accompanied the government to biosafety negotiations, working closely with the Global Industry Coalition, for example. Its promotional work is done through seminars, workshops, press work, and information dissemination, without engaging in direct political lobbying: doing the "backstage" work to complement the lobbying that companies do on their own behalf.[8] They do, nevertheless, engage in joint activities with the office of biotechnology within the Comisión Nacional Asesora de Biotecnología Agropecuaria (CONABIA), the main approval body for commercial applications of GM.

The key seed growers' association, ASA has a long history of representing seed-industry interests in policy debates. Formed in 1949, with sixty-four associated companies, it has come to represent a combination of national and foreign capital from firms such as Sursem and Satus to multinational giants such as Bayer, Dow AgroScience, and Monsanto. Indeed, the posts of president and vice president of the association are occupied by Dow and Monsanto respectively. ASA is also a member of other associations for the promotion of specific crops and, though less globalized than ArgenBIO, has ties to the International Seed Federation.

ASA operates as an industry bloc, having more direct interactions with government bodies such as the approval body CONABIA and the Servicio Nacional de Sanidad y Calidad Agroalimentaria (SENASA), the body charged with assessing food safety, where they are said to carry "substantial weight" and enjoy "good personal relations."[9] The number one issue for ASA has been intellectual property rights (IPRs), for which a separate committee exists within the body (ASA 2006). Within these associations, there are of course differences in corporate strategy that need to be reconciled or at least respected. The contentious question of technology fee payments for Roundup Ready (RR) soybeans has provided one focal point for diverging positions, with many ASA members wanting to avoid paying royalties to Monsanto, while the executive committee of the body is also bound to be responsive to one of its most powerful members and the company represented by its vice president, Monsanto.

A second divisive issue for the association was a government intervention that divided members along the lines of domestic and foreign capital. Resolution 71 in 2006 provided a waiver of ninety days of normal seed approval regulations for hybrids containing the GA21 event owned by Syngenta. The resolution gave the right to all seed companies to register their GA21 corn hybrids even if they did not "discover" the event or have a license from Syngenta to use it. According to one industry coalition spokesperson, the decision represented "a loss of credibility for the system as a whole,"[10] a view echoed by Monsanto and Dow, but seen as an opportune move by some national companies within the association. For an enterprise such as Atanor that had invested considerable sums in corn but without license from Syngenta, this provided an enormous advantage, with claims from the company that they would not be able to sell RR maize bags for half their current price. In response, Syngenta has threatened to contest these approvals legally. The strong reaction of ASA to this course of events, which many members were unhappy with, has further highlighted the fissures between domestic and foreign capital. This has now led to the creation of a rival seed growers' association, the CAS (Cámara Argentina de Semilleros),[11] entrenching the distinctive perspectives of foreign and national firms and, according to one rural media source, "disputing the representativity of a business worth USD$800 million" (InfoCampo 2006).

Given that each association has a different political profile in terms of public positions and connections with different parts of government reflecting its membership profile, some companies are members of several such bodies to maximize their influence through multiple channels. Particularly firms with interests in biotech and agrichemicals, such as Dow, are represented in ArgenBIO, ASA, FAB, and Cámara de Sanidad Agropecuaria y Fertilizantes (CASAFE).[12] While the former advances their interests in biotech, the latter represent their interests in seed markets and trade rules and chemical regulation respectively. Firms are also clear that some memberships in associations are maintained on a "nice to have" basis and others on a "need to have" basis.[13]

Membership in such associations performs several important functions for their members. One is about learning what other firms are doing, finding out about experiences they have had in the regulatory process, and building common positions as well as developing strategy. Though firms clearly compete with one another in the market and one company's innovations can have a negative impact on another, they confront common problems with regard to the regulatory process. For the purposes of registering their concerns on these issues, common fronts make strategic commercial and political sense.[14] This happens, for example, with regard to a novel food or technology event such as gene stacking, where new regulatory challenges arise that affect the sector as a whole. Around these issues, information sharing regarding scientific studies and previous relevant regulatory experience takes place, without disclosing commercially confidential information about one another's research lines or future market strategies. So even where there is competition over market share, there is a broader common political battle to secure acceptance of the technology with society and with the government, which transcends individual corporate strategies. Or as one regulatory affairs person put it, "We compete in the market but with respect to regulation we are partners, a team."[15]

At times, and with regard to sensitive issues that divide industry associations, such as IPRs, companies will fight their own battles, as is the case with Monsanto's ongoing conflict with the Argentine government, discussed further below. Some companies are also better equipped to go it alone politically than others. This is partly an issue of resources and capacity to contest the issue on all fronts. Monsanto, for example, is the only company in Argentina to have their own office of scientific affairs, which works for the whole region. Clara Rubenstein, a biologist who works in this office, represents ASA on the SENASA committee that deals with issues of food safety. In this sense there is a blurring of the boundaries of the firm, where an individual may be required to simultaneously represent a company, an association, and a sector.

Market position is, nevertheless, a strong determinant of policy positions. Firms such as Cargill have to meet the demands of food retailers in countries with hugely divergent product and process demands. Indeed, differences of opinion between industries with regard to the regulatory system in Argentina often reflect their status as producers or exporters. While producers wanted an end to the "mirror policy" once the EU de facto moratorium took hold, many exporters did not, fearful of loss of market access to the European Union and wanting to replace the United States as the provider of non-GM maize to Europe. These conflicts were played out in relation to approvals of maize T507 and GA 21 in roundtables overseen by Dirección Nacional de Mercados Agroalimentarios (DNMA), the body responsible for export markets, where producers pressured for the abandonment of the mirror policy, but exporter interests, which traditionally have a closer relationship with DNMA, were able to prevail in ensuring its (temporary) continuation. Once again, differences in interest between firms over particular approval decisions, or even more generally around the rules by which the technology should be governed, do not negate evidence of industrywide support for the technology and the model of agricultural development that sustains it.

Corporate Power in the Politics of Biotechnology in Argentina

This section of the chapter seeks to understand the sources of power exercised by agribiotech corporations in the contemporary politics of biotechnology in Argentina through reference to their material contributions to the Argentine economy in their capacity as key traders, exporters, and employees, their institutional presence in the deliberations of government on the issue of agricultural biotechnology and their contribution to the construction and maintenance of a particular discourse about biotechnology and development. It is important to note that while for the purposes of analytical clarity these dimensions of power are discussed separately, it is their interaction and mutually reinforcing nature that give these corporations such power in the contemporary political economy of Argentina. I have referred to this elsewhere in terms of the exercise of "bio-hegemony" (Newell, 2009).

The discussion of material power below is more closely related to the corporation's relationship to a particular structure of (agricultural) production from which it derives its power. The critical role of companies in the process of production and the wealth this generates (for themselves and indirectly to the state through employment and taxation) confers on corporations the structural power to advance their agenda, and explains the receptivity of state officials to business positions reinforced through well-resourced lobbying campaigns. Structural power then describes a *type* form of power, of which material power can be one source. For Strange (1988), for example, structural power describes the way A gets to shape the context in which B makes decisions. Through their material power, corporations in Argentina do, at times, exercise structural power, but it is the *source* rather than the type of power that is being described here. Hence, while the effect may be to constrain state autonomy, the focus here is explaining the source of that power, which in this case is material. Likewise, while in the introduction (chapter 1), instrumental power is used to describe some of the forms of lobbying and access to institutions that I analyze in terms of institutional power, I would argue that institutional access and power are at once sources of power that allow firms to assert their political objectives, but also manifestations of power that derive from their (material) contribution to the capital accumulation objectives of the state. It is that which affords them privileged channels of access to decision making. As with discourse, the third pillar of corporate power examined here, the power of that discourse is intimately related to the structure of production and corporate sponsorship of key media in this area. Positive discourses around the role of biotechnology in Argentine agriculture are privileged and given widespread attention because they lend support and credibility to key state and corporate accumulation strategies to the exclusion of more critical views. As noted above, although these sources of power are treated separately to explore their distinct dimensions, it is their interaction in mutually supportive ways that serves to ensure that the power of agribusiness in Argentina is currently hegemonic (Newell 2009), albeit not complete, and that challenges to the merits of biotechnology are rendered nonissues.

Material Power

Agricultural biotechnology clearly forms a central state strategy for Argentina. This was reaffirmed by the National Plan in 2005, but proactive support for the technology has a much longer history, as noted above. People often refer to the politics of agbiotech in Argentina as a "nonissue." With a strong alignment of the interests of state, national, and foreign capital about the value of biotechnology, the key contestations are around access and ownership of the technology, rather than around its desirability in social and environmental terms. Figures regarding the percentage of agriculture devoted to biotech and the percentage of exports based on agriculture, noted above, make it abundantly clear that the material contribution of the biotech sector to the Argentine economy is immense. Indeed, the very nature of the approval system, as we have seen, is structured around the export potential of the technology. The political implications of this structure of production are apparent through patterns of trade in which companies are key players and through the property rights they seek in order to guarantee a return on their investment (Newell 2003). The centrality of biotech companies at all stages of the production process in creating, cultivating, and exporting GM crops affords them a central role in the politics of the technology in Argentina. Property rights are a key means by which control over the seed is guaranteed, extending their power from production in the field to the consumption of food.

Trade The politics of biotechnology have to be understood in relation to Argentina's trade politics, within Latin America and globally. What happens in Argentina's key export markets is a significant determinant of the course of events within the country. The turn to China has been particularly notable in this regard. The warming of relations toward China has been gradually consolidated, as demonstrated by a state visit in 2004 in which a partial write-off of Argentina's debt in exchange for access to investments in key service sectors was considered, as well as by increasing agricultural trade links between the two countries. When the price of soy fell in Argentina, China looked to strengthen trading ties and Argentina was well placed to increase exports to such a huge market. Business groups such as the Cámara de la Producción, la Industria y el Comercio Argen-China[16] have a longer history of trying to increase trade ties between the two countries, active since 1984. Not all firms have welcomed closer ties, however. In an effort to protect its position, Monsanto, for example, pressured the Argentine government to initiate a dumping case against China following the alleged dumping of imports (glysophate). In February 2004 the government decided not to pursue the case, a decision that found support among the agricultural sector as a whole, which was worried about antagonizing a key trading partner.

There is an interesting regional dimension to the trade politics of biotechnology. This derives from the recognized, but illegal, flow of GM seeds (or "Soja Maradona," as it is dubbed) between Argentina and Brazil via Paraguay, which has led to representations by the Brazilian government to Argentina requesting tougher measures to control this unregulated diffusion of GM seeds. Argentina claims the seeds are meant for processing as oil in Paraguay, but are illegally bagged and sold for direct growing in Paraguay and Brazil.[17] Government officials also, however, acknowledge the near impossibility of effectively regulating the trade in a region with so many shared borders in such close proximity.[18] Others suggest the use of the seeds for cultivation rather than processing is with the direct support of biotech companies, for whom it is a "logical firm strategy," as one interviewee put it. There is evidence indeed of farmers from neighboring countries being offered seeds at rural fairs that were not approved outside of Argentina. Argentina in this sense was seen as a useful platform from which to penetrate markets in Brazil and Paraguay, according to some government officials. We can see with this example, then, the way in which, particularly concerning a sector whose products can pass easily across borders and between traders without state authorization, corporations can easily bypass formal regulations in seeking to access and penetrate new markets for food and agricultural produce.

Within the institutions of Mercosur, such as the SGT6 Working Group on Environment, despite early framings regarding the potential risks associated with the technology, Argentina played a lead role in vetoing the biosafety clause of a proposed draft. The Framework Agreement accepted in 2001 has no section at all on biosafety issues (Hochstetler 2003, 2005). When Argentina called a meeting of Ministers of Agriculture in Mercosur in 2005 to generate support for its position against paying Monsanto royalties on soy crops (rather than seeds), initial support was forthcoming from Brazil and Paraguay. Fierce pressure in the wake of the meeting, however, led to these governments retracting their positions on the basis that they were concluding their own agreements between the private sector and Monsanto. Intense lobbying by Monsanto on those governments not to prejudice their own bargaining position by declaring support for the Argentine position was assumed by Argentine officials to be behind this aboutface. These interactions appear to provide evidence of what Susan Strange (1994) calls "triangular diplomacy" where states are negotiating with one another, with firms operating in and beyond their own borders, all alongside interfirm bargaining over market access and policy positions.

An intriguing element of the *global* trade story is the case brought before the WTO dispute settlement panel by the United States and its allies, including Argentina, contesting the EU de facto moratorium on the commercial approval of GM crops. Government officials concede that Argentina was under considerable pressure to sign up to the case and to demonstrate solidarity with their larger allies in the Miami group of GM exporting nations.[19] Some elements within government were reluctant to launch an offensive against the EU, which Argentina is dependent on for trade and with whom it had managed to avoid trade conflicts through its mirror policy of only approving for commercialization crops that had already been approved in the country's key export markets. Though companies such as Monsanto, following the line of their headquarters in St. Louis, were keen to have the case brought, many firms in Argentina were weary of the merits of bringing the case. Grain and food traders such as Cargill were also reluctant given the significance of the European market for their products.[20]

The difference in approach toward the case among corporations is illustrative of different styles of policy engagement. Monsanto's ongoing conflict with the Argentine government over IPR protection for its products, which resulted in the US giant trying to collect fees in European ports where products produced from seeds in Argentina were arriving, is a case in point. Within the industry coalitions in Argentina, there is recognition that Monsanto is more aggressive in its pursuit of its aims and more vocal, combative, and open about its confrontation than most Argentine firms, which are more accustomed to lobbying discretely and behind the scenes.[21] This difference in lobbying style between companies, often reflective of their home base, has been noted elsewhere (Coen 2005). The concern on the part of other elements of the biotech industry was the potential for Monsanto's more conflictual strategy to sour relations and strategic advantages for the sector as a whole.[22]

What is also interesting is the way the state had to steer policy toward reconciling the needs of "capital in general" with overall state strategy (Holloway and Picciotto 1978). A leading figure in the approval body CONABIA, who was one of a team of four representatives[23] from Argentina that attended the hearings in Geneva, expressed his frustration in relation to the WTO case. He complained that biotech firms were "thinking in the very short term, product by product" and losing sight of the need to win the overall battle to secure access for biotechnology products to the European market, to "defend the long term political strategy that is key for Argentina's national development."[24] In this sense, the positive verdict of the WTO dispute settlement panel is seen as "symbolic,"[25] important in the longer-term perspective of securing political conditions conducive to the technology's development, but unlikely to lead to immediate retaliatory action on the part of the parties that bought the case because of the dependency on European markets described above.

Property Rights IPRs are argued by firms themselves to be key to their ability to control access to their products, yet in Monsanto's case, the fact that the firm does not have an exclusive patent for its seeds has generated an intense conflict with the government of Argentina. Monsanto licensed the Argentine firm Asgrow Argentina to have access to its Roundup Ready gene. When the multinational firm Nidera acquired

Asgrow in the late 1980s and with it access to the gene and the right to use all Asgrow's germplasm, it became possible to disseminate the seed widely in Argentina. In the mid-1990s Monsanto bought Asgrow International grain and oilseed business, ending the free-access agreement with Nidera for new breeding lines, but Nidera kept control of existing lines. In 1996, therefore, commercial authorization was granted to Nidera for RR soybeans, the seeds of which were sold to farmers without purchase contracts.

When Monsanto requested an exclusive patent on RR soybean seed, it was denied on the grounds that the gene had already been released with many plants expressing the RR gene and hence was no longer "novel."[26] Since Argentina adheres to the 1978 version of the Union for the Protection of New Varieties of Plants (UPOV) treaty, which allows farmers to save seed for their own purposes,[27] and since the herbicide-resistant gene in RR soy comes from the seed, farmers are able to reseed without paying royalties to Monsanto. This has given rise to a huge black market in "bolsa blanca" soybean seeds, which by 1997 covered 68 percent of the market and that keep the price of RR soybean seed in Argentina below global market prices. Today the figure for illegal seeds planted in Argentina is said to have risen to over 80 percent.

There is a great deal at stake, economically and legally: US$300 million per year according to one estimate.[28] Monsanto has sought to sign contracts with farmers regarding access to the technology, exerting pressure on them to sign agreements couched in technical and legal terms that are alien to most farmers. Despite government support in interpreting and explaining the agreements, as one IPR lawyer for the Ministry put it, smaller producers "really have no choice but to sign."[29] National firms such as Bio Sidus have also been pressuring the government to prevent abuses of "farmer privilege" provisions by protecting their products.[30] Around the issue of property rights protection there is common ground with Monsanto and other TNCs where policy positions can be advocated by the Foro Argentino de Biotecnología which represents both national and foreign firms.

Also interesting is the transnational element to this dispute. The conflict over IPRs between the Argentine state and Monsanto was driven, in part, by producer interests in the United States who were disgruntled with the fact that they were having to pay technology fees to use Monsanto products while farmers in Argentina were apparently getting away without paying. This was keeping the overall costs of production lower and reducing the price of soy and other products on the international markets, making them more competitive than their counterparts in the United States. One interviewee suggested in this regard that the powerful US corn growers' and soybean associations[31] were pressuring Monsanto to toughen its stance toward the illegal seed sales in Argentina. Groups such as the American Soybean Association (ASA) have raised these issues in the past in relation to Brazil. A press release from ASA reads as follows: "ASA President Ford has taken action to ensure that US soybean farmers aren't being placed at a competitive disadvantage in this particular trade due to the fact that US farmers pay royalties for soybean varieties containing Roundup Ready technology while Brazilian farmers are reported to have widely pirated Roundup Ready seeds." The association head continues, "Ever since this contract was announced in May, ASA and the United Soybean Board have been working closely with Monsanto, which holds the patents on the Roundup Ready technology, to ensure that the imported soymeal is not derived from pirated Roundup Ready soybeans. . . . Monsanto has assured us that it will enforce its intellectual property rights regarding all importations of Brazilian soybean meal or soybeans" (ASA 2002).

There has been significant pressure to strengthen national systems of patent protection from both the pharmaceutical and agribiotechnology industry, which has succeeded in bringing about changes favorable to industry.[32] Political pressure and vast sums of money were mobilized toward pushing the Argentine Congress to loosen restrictions on what can be patented and by whom.[33] More sinister are the repeated approaches that have been made to government officials to come and work for Monsanto.[34] Given the discrepancy between salaries paid by TNCs and those earned by public officials in Argentina, such offers are a great temptation. For the company, having an employee with direct contacts and inside information about government strategy and operations is invaluable, though they also run the risk of alienating the government they are trying to outmaneuver by using what one official referred to as "dirty methods."

As has happened in many other countries such as India and China (Newell 2008a), industry lobbying is backed by US government pressure, with Argentina appearing on the US government IPR "watch-list" of countries allegedly not fulfilling their Trade Related Intellectual Property Rights (TRIPS) obligations. This intercapital conflict among producers has created interstate tensions, with the US Secretary of State intervening on behalf of Monsanto to make representations before the Argentine Secretary of Agriculture, Miguel Campos, about IPR protection. Campos suggested that the key alliance that exists between the United States and Argentina as part of the Miami group in the biosafety negotiations, and as partner in the WTO case, was at put at stake by such overt gestures on behalf of the company.

In December 2003 Monsanto stopped selling its own seed in Argentina, and in 2004 it terminated soybean research and marketing, a strategy aimed at pushing the government into negotiations about compensation and technology fees. A bill promised at the end of 2004 did not materialize following strong resistance from farmers' groups such as Federación Agraria Argentina, which represents a broad cross-section of over 100,0000 smaller producers. Monsanto's response to the legislative stalemate has been to attempt to collect royalties by other means. Shifting venues in June 2005, Monsanto filed lawsuits in Denmark, Spain, and the Netherlands, where it does enjoy patent protection, in order to collect royalties on imported RR soy from Argentina. Monsanto was able temporarily to block the shipments from entering European territory, from which both Argentine exporters and European importers of soymeal suffered extensive losses, including legal expenses and costs of storing the merchandise delayed at the ports (Biadgleng and Tellez 2008). A court in Madrid ruled in September 2007 that the Spanish importer Sesostris, controlled by the international commodities group Louis Dreyfus, was exempt from such payments (Misculin 2007), but Monsanto has vowed to appeal the ruling. The government feels itself to be on the front line of a broader political contest over the appropriate boundaries of patent protection, because for the company to claim ownership of both *product* and *process* implies a level of consolidated control over the production chain that has far-reaching repercussions for government and industry the world over. For the government, the contest centers on whether a company has the right to "appropriate everything," "to extend their control from the seed to the supermarket," as one official put it.

Given these stakes, there has been a fascinating realignment in the fault lines of conflict. Other governments, notably even those with which Argentina is in conflict within the WTO, have offered legal and technical support, and even traditional antibiotech opponents such as Greenpeace have volunteered their services.[35] There is a sense on the part of the government that Monsanto's strategy is one of attrition: to draw out the legal cases since they have the funds and personnel to outlast the resistance of the government. The cases could take at least five years to conclude in total (including prehearings, the court case, and appeals) and Monsanto has access to the very best lawyers to support their claims to patent protection.[36] In this sense, the cases appear to lend support to Carlos Correa's (2006) claim that IPRs provide a new instrument of control for TNCs over developing countries, which are often powerless to challenge them amid such disparities in resources. Forum shifting and using legal remedies available in other countries' jurisdictions is a key manifestation of corporate power: a way to outmaneuver underresourced governments (see Sell, chapter 7, this volume).

This confrontational approach, at odds even with other producers and biotech firms, has served to antagonize the Argentine government, provoking strong responses. In June 2005 Agriculture Secretary Miguel Campos told a press conference that "Monsanto has shown that it continues to be a national embarrassment" (Nellen-Stucky and Meienberg 2006). The government maintains that Monsanto's position is anticompetitive, abusive of their market position, and contradictory in the sense that it has stated publicly in the past that RR soy and conventional soy are "substantially equivalent," undermining the current patent-related claim to novelty made of the technology. Industry groups, meanwhile, have been critical of the way the discourse of national sovereignty over seeds adopted by the government has been invoked amid references to "our capital," counterclaiming that this is foreign and not Argentine.[37]

Institutional Power

Biotech corporations are heavily involved in formal decision making around biotechnology in Argentina. This is hardly surprising given their material contribution to the economy, described above. In their own right, as well as through associations such as ASA and FAB, they seek to ensure their voice is heard in discussions about the formulation and implementation of policy. Their close links with government have led to accusations of a revolving door, or more severely of co-optation of leading ministers by biotech interests. Greenpeace Argentina (2004), for example, went so far as to label former Minister of the Economy, Roberto Lavagna, "employee of the month of Monsanto." The fact that Lavagna founded and worked for the firm ECOLATINA, which was hired by Monsanto during the trade dispute with China mentioned earlier, only serves to fuel the activist claims.

Apart from their formal participation in key decision-making bodies, discussed below, public and regulatory affairs personnel from biotech companies claim to meet with government officials of different ministries, depending on the issue, every two to three weeks. Changes to the regulatory system are discussed at length, involving considerable discussion with industry representatives before key officials make a final decision. Likewise, before key international meetings and in their wake, by way of feedback, consultations are held with companies depending on the issue under discussion. There was extensive formal and informal consultation, for instance, with biotech corporations and seed traders prior to and during the WTO dispute with the EU.[38] There are also meetings with associations such as ASA prior to WTO meetings or negotiations around plant genetic resources in an attempt to build consensus positions on key issues.[39] The nature of the relationship differs by government department, and the cycles of interaction are determined by broader patterns of political events, such that meetings between biotech business and government were more intense in the years 2000 and 2001 in the wake of global controversies around the technology and then in the runup to the WTO panel case in 2006.

CONABIA is, in many ways, the epicenter of the approval process for agricultural biotechnology applications. It is a multisectoral body, wherein private and public organizations are represented.[40] Membership and coordination have been modified with increasing input from the private sector. Currently, CONABIA consists of three public research institutions, four public universities, six private-sector associations, one civil society (consumer) organization, four representatives from the Secretaría de Agricultura, Ganadería, Pesca y Alimentos (SAGPyA) offices, and two from the Health Ministry, though this composition is subject to change over time. Companies participate through chambers and associations such as FAB and ASA rather than as individual firms, often adopting common positions on the issues under discussion. Nevertheless, individuals representing those bodies come from leading firms such as Monsanto, Syngenta, Dow and Bayer.

Given the range of government institutional actors that have a role to play in the governance of agricultural biotechnology in Argentina, it is unsurprising that some firms have closer ties with some committees, departments, and ministries than others. Different parts of the state also have distinct regulatory responsibilities, so that on questions regarding the implementation of the Cartagena Protocol (which Argentina has signed but not ratified) or labeling, for example, firms work closely with the Cancillería (Ministry of Foreign Affairs), while the seed law requires close cooperation with INASE (Instituto Nacional de Semillas)[41] and the Ministry of Agriculture. The Ministry of Agriculture is described by regulatory affairs staff of biotech firms as being very "pro-technology" and "pro-production" and in this sense is very responsive to their views. Despite good relations with the Cancillería, the support of this ministry for the ratification of the Cartagena Protocol, in opposition to the Ministry of Agriculture, drew fire from industry. In reality, though, there is not much dispute around the Protocol since Argentina has not ratified it and most officials regard it as more or less irrelevant to the day-to-day trade in GMOs. Predictably, the pattern of interaction goes beyond the distinct positions of government departments and the competing agenda of industry lobbies to personal relationships between government and industry personnel within and beyond formal decision-making committees.[42] For example, the Secretary of Agriculture appointed in 1999 was more precautionary in his outlook toward the trade consequences of biotech approvals, while Roberto Lavagna as Minister of Economics under the administration of Kirchner, perhaps for the reasons given above, has been more strident.

Different parts of government also have distinct cultures of engagement with the private sector. Within INASE, the agency responsible for seed registering and commercialization as well as plant-variety protection, specific industry roles are defined by regulations. These require seats on the board that directs the body for five private-sector representatives, including one for seed producers, one for phytoimprovers, one for plant-variety holders, and two for seed buyers. Conveying the sense of an equal partnership between the public and private sectors in seed-market regulation, public-sector officials are entitled only to the equivalent five seats on the board.

Where more formal systems of private-sector representation are absent such as exist in INASE or CONABIA, more informal dialogues, roundtables, and exchanges with industry are commonplace. This happens, for example, within DNMA, the body responsible for export markets. DNMA has input into decision making about the commercial potential of particular applications[43] and contact with the private sector is "constant and iterative," according to DNMA officials. Corporations are encouraged to submit comments and studies related to their exports. Access to commercial information is critical and corporations are obviously key "street-level bureaucrats" in this sense. Again, there are important differences in power and access, reflective of market share, between Cargill and other multinationals on the one hand, and the small producers, exporters and the broader associations of seed growers and farming interests such as Asociación de Cooperativas Argentinas on the other.

Food safety aspects of the technology (including toxicological effects, allergenicity and nutritional value) are governed by SENASA. This organization has a technical assessment committee regarding the use of GMOs that receives reports and studies produced by companies on these issues, including previous information submitted as part of approvals in other countries. As with CONABIA, researchers represented on the technical assessment committee are meant to be independent of the private sector applications they are evaluating, though officials concede that some researchers inevitably do work for companies whose applications are being assessed.[44] Applications are judged according to principles of substantial equivalence and methodologies promoted by the WTO–World Health Organization (WHO) Codex Alimentarius, whose task force has an Argentine representative from SENASA. Indeed, several passages of Resolution 412/02 "are directly lifted from the relevant Codex documents" (Van Zwanenberg 2006, 20). This resolution was drawn up by a committee consisting of government officials, including members from CONABIA and the biotech and related industries, represented by groups such as FAB.

Here, then, we can see a clear role for companies in "domesticating" global policy and shaping the terms of implementation of international standards (Newell 2008a). What is even more interesting is the extent to which the same TNCs are heavily implicated in shaping Codex standards (Smythe, chapter 4, this volume). Corporate power exercised at the international level, therefore, circumscribes the policy autonomy available to countries at the national level. Autonomy to make use of national "policy space" permitted within often loosely worded agreements can, in turn, be restricted by lobbying and veto roles performed by companies, which will often be expected to implement the regulations and fund the studies to prove that they have done so. Therefore, the global reach of multinational companies and their embeddedness within global policy networks, means that not only are they better placed to outmaneuver commercial rivals, but they can also outmaneuver under-resourced national governments that can often not afford to take part in and effectively shape global standard-setting processes.

Just as important, however, are the informal networks and exchanges that take place between company staff and members of government. "Charlas previas" (prior conversations) and dialogues with government are available to those associations that are not directly represented in the key decision-making bodies. Events hosted by the organization ArgenBio, part of the global network of "Bio" industry associations formed to advance the interests of biotech multinationals, are important in this regard. Seminars, often on very technical themes concerning biotech, provide an opportunity for people to come together, exchange gossip about the latest technological and political developments and changes of personnel, and seed new policy ideas.

Discursive Power

The third source of corporate power that is key to understanding the nature of the politics of agricultural biotechnology in Argentina is discursive power (see Williams, chapter 6, this volume). The access to and ability to sponsor influential mass media in the country plays a crucial role in generating and maintaining support for biotechnology and denying space to critical or dissenting voices. Media-production routines and hierarchies of what counts as valid expertise reinforce business framings of biotechnology and its benefits for Argentina. In a country where *el campo* ("the countryside") generates powerful cultural resonances and is central to the nation's identity, rural issues are guaranteed a high place on the political agenda. Those able to promise gains in rural incomes, especially in the aftermath of a major financial crisis, are seen as national saviors rather than as purveyors of a risky and untried technology, as they sometimes are in Europe, for example. There is a high level of receptivity, therefore, to a media and public discourse proclaiming the benefits of agricultural biotechnology.

The hegemonic discourse in Argentina regarding agricultural biotechnology is that it represents an important, economically significant, socially beneficial, safe, and environmentally benign technology (Newell 2009). This frame is sustained through government speeches and policy documents (SAGPyA 2005), the publicity work of individual companies and associations through seminars, conferences, press conferences, and constant advertising in the media aimed at policy and public audiences, as well as through billboards in the countryside aimed at reaching farmers directly.

In terms of the mainstream media in Argentina, there are very few dissenting voices challenging the benefits of agricultural biotechnology. Industry coalitions concede that the media is both "supportive" and "sensitive" to their positions.[45] The left-leaning critical *Página 12* newspaper and its associated magazine *Veintitres* have published articles that depart from the consensus, but these reach a relatively small proportion of the public. Likewise, newspapers like *La Tierra*—put out by producer groups such as Federación Agraria Argentina—adopt a skeptical position on biotech but do not reach large audiences. *La Nación* and *Clarín*, the country's two most widely read newspapers, on the other hand, are strongly supportive of the technology. *La Nación* publishes a "Campo" section every Saturday that serves as a forum for issues of concern to the agri-industry and *Clarín* does the same with its "Rural" section, also published every weekend. Heavy levels of advertising sponsorship from key agricultural producers help to ensure a media responsive to their concerns. Each "Campo" and "Rural" section is full of large advertisements from Bayer, Monsanto, and Syngenta, as well as a wide range of other agrichemical producers proclaiming the benefits of "super-soja" with a Superman soy pellet used to reinforce the point. The relationship between advertiser and client is reciprocated through sponsorship by *La Nación* and *Clarín* of the annual "ExpoAgro" agribusiness exhibition and trade fair. Photos taken at such events and featured in these supplements show senior government officials, heads of companies like Monsanto, and leading journalists from the newspaper lined up next to each other. This practice allows the newspaper to demonstrate both its proximity to the centers of power and support for the agricultural sector whose profits sustain its advertising stream.

Argentina's influential *Sociedad Rural*, a wealthy producers' organization, hosts an annual show that attracts agricultural interests from all over the country. Key leaders of farming associations, together with politicians keen to align themselves with powerful rural lobbies, give speeches that articulate key demands regarding support from government in the form of tax concessions and the like, which are faithfully reported in *La Nación* the following day. President Cristina Kirchner is currently under attack from what she has dubbed the "4 by 4" large landowning farming lobby (in reference to the cars they drive) for raising the level of taxation on exports of crops, including (GM) soy, whose prices have recently soared. Early indications are that the blockades and forms of protest adopted by the farmers, usually condemned when applied by the left-wing *piqueteros* movement of the unemployed, are viewed sympathetically and the cause a just one by the right-wing press (Newell 2009). More pro-government media such as *Clarín* are left in an awkward position, sympathetic to the plight of the farmers but also obliged to extensively report the government line on the crisis.

In terms of media production and the sourcing of news material, there is a hierarchy in terms of who the press turns to for information about biotech-related issues in Argentina. For news of the latest scientific developments or to write opinion pieces, the media often come to groups such as ArgenBIO, which, as noted, operate as publicity outfits for multinational biotech firms. This is true of the mainstream daily newspapers, as well as of specialized magazines such as *Nuestro Campo* (Our Countryside) and *Ciencia hoy* (Science Today). In general, therefore, despite minimal coverage of more critical positions in elements of the press with less circulation, the mainstream media help to ensure that biotechnology remains a "nonissue" in Argentina.

Conclusion

The politics of agricultural biotechnology in Argentina present a series of fascinating insights into the role of agrifood corporations in the global economy. Biotechnology corporations in Argentina operate in an economy heavily reliant on agriculture, yet with strong ties to markets such as Europe where resistance to biotechnology is deeply embedded. They are subject to a regulatory system that is unique in its explicit evaluation of GM crops according to their ability to create significant export earnings, but weakly institutionalized in terms of the legal weight that the resolutions carry. They are key actors in the sense of acting as the "street-level" bureaucrats of biotechnology regulation, managing the day-to-day governance of seeds and biosafety, as well as being deeply involved in the formal processes of regulation at the national level, including the formulation of regional and international agricultural trade strategies. Agrifood corporations in Argentina operate in an environment in which a strong state commitment to biotechnology exists alongside an almost total absence of contestation around their role in the economy or the value of the products they produce, even if disputes continue about the means of protecting them. And yet still, through ties to other markets and through links to colleagues in the same firms but operating overseas, they are sensitized to the controversies surrounding the technology.

The different roles that firms play in the process reflect the diverse degrees of material power they wield, differential levels of institutional access, and divergent intensities of engagement in public-political debate about the role of biotechnology in the future of agriculture in Argentina. There appears to be an intimate link between the corporate strategies of firms and the public-political positions they adopt, and, therefore the nature of their engagement with the state. At times their power is exercised in purely commercial terms, through their role as seed traders, employers, and street-level bureaucrats providing information critical to decision making and enforcing (or not) market regulations set by government. Through these means, they are central actors in the governance of biotechnology. In more formal governance roles, they serve as advisors, regulators, and representatives of government when they sit on national delegations. As political actors in the market and with and beyond the state, firms in all parts of the biotech chain, though some more than others, will remain central to the political economy of biotechnology in Argentina.

In an environment such as this, we confront the limits of thinking critically about the implications of a corporate and multinational company-led model of agriculture. With material, institutional, and discursive sources and forms of power so closely aligned, the policy space to reflect on the range of social and environment impacts being generated by this developmental path is just not there. Though positive claims are made of the technology's potential to benefit the poor and reduce the environmental impacts of intensive agricultural development, fuller and more open forms of deliberation and public engagement about the technology and what might constitute appropriate forms of social control of it, are almost impossible to envisage. If concerns about the sustainability of this trajectory of agricultural development are to register on the radar of government and business elites, it will more likely be via strong market signals sent down supply chains or through other governments' trade restrictions, rather than from within the country that some activists have dubbed "la Republica Unida de la Soja."[46]

Acknowledgments

I am grateful to comments and suggestions for improvement from the editors Jennifer Clapp and Doris Fuchs, the other contributors to this volume, and to the anonymous reviewers.

Notes

1. In many cases the identity of interviewees is protected to ensure anonymity, given the sensitive and controversial nature of some of the interview material.

2. The extent of deforestation that can be attributed to the expansion of soy production is disputed. See for example Brown et al. 2005.

3. See www.conabia.sagpya.gov.ar.

4. Association of Argentine Seed-Growers.

5. The Argentine Biotechnology Forum.

6. Association of No-till Producers.

7. Interview with Executive Director, ArgenBIO, October 20, 2006.

8. Interview with Executive Director, ArgenBIO, October 20, 2006.

9. Interview, industry organization, October 2006.

10. Interview, industry association, October 2006.

11. Argentine Chamber of Seed Growers.

12. Chamber of Farming Health and Fertilizers.

13. Interview with regulatory affairs official, TNC, October 2006.

14. Interview with Dow Agrosciences, October 2006.

15. Interview with Dow Agrosciences, October 2006.

16. Argentine-China Chamber of Production, Industry and Trade.

17. Interview with Mónica L. Pequeño Araujo, Coordinadora de Proyectos Especiales en Biotecnología, Instituto Nacional de Semillas (National Institute of Seeds), October 24, 2006.

18. Interview with official from Ministry of Agriculture, November 2006.

19. Interestingly, though the United States was active in pushing the case in the first place, Canada and Argentina were more active participants during the actual hearings. Interview with representative from Argentine delegation, November 2006.

20. Interview with grain trader, Cargill, October 2006.

21. Interview with head of industry association, October 2006.

22. Interview with head of regulatory affairs, TNC, October 2006.

23. The delegation included two lawyers and two technical specialists. Interview material.

24. Interview at CONABIA, November 2006.

25. Interview with biotech sector analyst, October 26, 2006.

26. Interview with Maria Laura Villa Mayor, lawyer, INASE, October 24, 2006.

27. It also makes no distinction between GM and non-GM varieties.

28. Interview with Eduardo Trigo, October 26, 2006.

29. Interview with IPR lawyer, Ministry of Agriculture, November 10, 2006.

30. Interview with IPR lawyer, Ministry of Agriculture, November 10, 2006.

31. This would certainly be consistent with broader America Soybean Association strategies that are targeted at biotech policy in developing countries. The association's website notes: "The US State Department has funding to support programs designed to influence key developing countries to support biotechnology in Codex negotiations and in continuing Biosafety Protocol implementation and negotiations. ASA and other ABPC organizations are working with the State Department, USDA, and other agencies to identify these countries, develop appropriate strategies, and implement coordinated public/private sector efforts to achieve positive outcomes." http://www.soygrowers.com/hot/hot10.htm.

32. Previous attempts to strengthen patent legislation in Argentina in line with TRIPs had been resisted by the domestic generic drug manufacturers.

33. Interview with IPR lawyer, Ministry of Agriculture, November 10, 2006.

34. Interview material, Secretary of Agriculture, November 2006.

35. Interview with official from Secretary of Agriculture, November 2006.

36. Interview with official from Secretary of Agriculture, November 2006.

37. Interview with head of leading industry coalition, October 2006.

38. Interview with personnel from Cargill, October 2006; interview with staff from DNMA, October 2006.

39. Interview with Mónica L. Pequeño Araujo, Coordinadora de Proyectos Especiales en Biotecnología, Instituto Nacional de Semillas (Nacional Institute of Seeds), October 24, 2006.

40. Each organization is represented by two persons, one titular and a substitute (SENASA is the exception, because it has four representatives, two specialized in animals and two specialized in plants; they alternate depending on the issue under discussion). These two are selected by the Secretary of Agriculture from three candidates proposed by each organization.

41. National Institute of Seeds.

42. Interview with senior regulatory affairs official of a TNC, October 2006.

43. Interview with Ruben Ciani and Federico Alais, DNMA, Seccion de Comercio Internacional, October 26, 2006.

44. Interview, senior official, SENASA, November 2006.

45. Interview with leading industry association, October 2006.

46. Adolfo Boy, Grupo Reflexión Rural, presentation at EU-Latin American trade summit, activist meeting, Vienna, May 2006.

References

Amoore, Louise. 2000. International Political Economy and the Contested Firm. *New Political Economy* 5 (2): 183–204.

Andrée, Peter. 2005. The Genetic Engineering Revolution in Agriculture and Food: Strategies of the "Biotech Bloc." In David Levy and Peter Newell, eds., *The Business of Global Environmental Governance*, 135–166. Cambridge, MA: MIT Press.

Asociación Semilleros Argentinos (ASA). 2002. South American Soymeal Imports Frustrate US Farmers. *Press Release*. August 30.

Asociación Semilleros Argentinos (ASA). 2006. Main web page. http://www.asa.org.ar.

Biadgleng, Ernius Tekeste, and Viviana Munoz Tellez. 2008. The Changing Structure and Governance of Intellectual Property Right Enforcement. *Research Papers*, No. 15. January. Geneva: South Centre.

Brown, J. Christopher, Matthew Koeppe, Benjamin Coles, and Kevin P. Price. 2005. Soybean Production and Conversion of Tropical Forest in the Brazilian Amazon: The Case of Vilhena, Rondônia. *Ambio* 34 (6): 462–469.

Coen, David. 2005. Environmental and Business Lobbying Alliances in Europe: Learning from Washington? In David Levy and Peter Newell, eds., *The Business of Global Environmental Governance*, 197–220. Cambridge MA: MIT Press.

Correa, Carlos. 2006. Monsanto v Argentina. *Le Monde Diplomatique* 82 (April): 4–5.

Falkner, Robert. 2007. *Business Power and Conflict in International Environmental Politics*. Basingstoke: Palgrave Macmillan.

Galli, Emiliano. 2005. De la Chaucha de Soja al Reactor Nuclear de Investigación. *La Nación*, Comercio Exterior Section, January 4, 2.

Garcia-Johnson, Ronie. 2000. *Exporting Environmentalism*. Cambridge, MA: MIT Press.

Glover, Dominic. 2007. Monsanto and Smallholder Farmers: A Case Study in CSR. *Third World Quarterly* 28 (4): 851–867.

Glover, Dominic, and Peter Newell. 2004. Business and Biotechnology: Regulation of GM Crops and the Politics of Influence. In Kees Jansen and Sietze Vellema, eds., *Agribusiness and Society: Corporate Responses to Environmentalism, Market Opportunities and Public Regulation*, 200–231. London: Zed Books.

Greenpeace. 2006. *Desmontes S.A.: Quiénes Están Detrás de la Destrucción de los Últimos Bosques Nativos de la Argentina*. Greenpeace Argentina. www.greenpeace.com.ar.

Greenpeace Argentina. 2004. Lavagna, el Empleado del mes de Monsanto. July 13.

http://www.greenpeace.org/argentina/bosques/lavagna-el-empleado-del-mes-d.

Hochstetler, Kathryn. 2003. Fading Green? Environmental Politics in the Mercosur Free Trade Agreement. *Latin American Politics and Society* 45 (4): 1–33.

Hochstetler, Kathryn. 2005. The Multilevel Governance of GM Food in Mercosur. In Robert Falkner, ed., *The International Politics of Genetically Modified Food*, 157–173. Basingstoke: Palgrave Macmillan.

Holloway, John, and Sol Picciotto, eds. 1978. *State and Capital: A Marxist Debate*. London: Edward Arnold Press.

InfoCampo. 2006. Lanzaron una nueva Cámara Semillera. November 3, 24–30.

Levy, David, and Peter Newell. 2002. Business Strategy and International Environmental Governance: Toward a Neo-Gramscian Synthesis. *Global Environmental Politics* 3 (4): 84–101.

Levy, David, and Peter Newell, eds. 2005. *The Business of Global Environmental Governance*. Cambridge, MA: MIT Press.

Misculin, Nicolas. 2007. Monsanto Loses Spanish Court Case on Argentine Soy. *Reuters*, September 7.

Nellen-Stucky, Rachel, and François Meienberg. 2006. Harvesting Royalties for Sowing Dissent? Monsanto's Campaign against Argentina's Patent Policy. Berne Declaration.

Newell, Peter. 2003. Globalization and the Governance of Biotechnology. *Global Environmental Politics* 3 (2): 56–72.

Newell, Peter. 2006. Corporate Power and Bounded Autonomy in the Global Politics of Biotechnology. In Robert Falkner, ed., *The International Politics of Genetically Modified Food*, 67–84. Basingstoke: Palgrave Macmillan.

Newell, Peter. 2007. Biotech Firms, Biotech Politics: Negotiating GMOs in India. *Journal of Environment and Development* 16 (2): 183–206.

Newell, Peter. 2008a. Lost in Translation? Domesticating Global Policy on GMOs: Comparing India and China. *Global Society* 22 (1): 115–136.

Newell, Peter. 2008b. Trade and Biotechnology in Latin America: Democratization, Contestation and the Politics of Mobilization. *Journal of Agrarian Change* 8 (2–3): 345–376.

Newell, Peter. 2008c. The Political Economy of Global Environmental Governance. *Review of International Studies* 34: 507–529.

Newell, Peter. 2009. Bio-hegemony: The Political Economy of Agricultural Biotechnology in Argentina. *Journal of Latin American Studies* 41 (1).

SAGPyA. 2005. *Introducción a las Negociaciones Internacionales en Biotecnología Agropecuaria*. Buenos Aires: Oficina de Biotecnología.

Saurin, Julian. 1997. Organizing Hunger: The Global Organization of Famines and Feasts. In Caroline Thomas and Peter Wilkin, eds., *Globalisation and the South*, 106–123. London: Macmillan.

Strange, Susan. 1988. *States and Markets: An Introduction to International Political Economy*. London: Pinter.

Strange, Susan. 1994. Rethinking Structural Change in the International Political Economy: States, Firms and Diplomacy. In Richard Stubbs and Geoffrey R. D. Underhill, eds., *Political Economy and the Changing Global Order*, 103–116. Basingstoke: Macmillan.

Van Zwanenburg, Patrick. 2006. *Risk Assessment Policies: Differences across Jurisdictions*. Argentina Case Study for ESTO Risk Assessment Policy Project, Draft, April.

Vara, Ana María. 2004. Transgénicos en la Argentina. Más allá del *Boom* de la Soja. *Revista Iberoamericana de Ciencia, Tecnología y Sociedad* 1 (3): 101–30.

Vara, Ana María. 2005. Argentina GM nation: Chances and Opportunities. NYU Project on International GMO Regulatory Conflicts. New York: New York University.

10

Corporate Power and Global Agrifood Governance: Lessons Learned

Doris Fuchs and Jennifer Clapp

This book has shown that corporations have come to play a key role in the setting of the rules and regulations that govern the global agrifood system. They establish private standards, participate in public-private partnerships, lobby national governments as well as multilateral institutions, and shape discourse on norms relating to issues such as health, sustainability, and risk. With this plethora of often-simultaneous activities, corporations have become major political actors in the global governance of agriculture and food. With the dramatic expansion of this political role of corporations and, in particular, of private governance institutions, the study of its implications for society is extremely important. As the case studies in this book have shown, the implications of corporate power in agrifood governance matter not only for the sustainability of the global food system, but also more broadly. After all, food is the basis of our existence and has pivotal health effects. At the same time, it provides an important source of livelihood for a large share of the global population and determines a wide variety of characteristics of the quality and well-being of local, regional, and global ecosystems.

Keeping this broader significance in mind, this book and the group of case studies included in it have pursued two goals. First, they have aimed at presenting a mosaic of corporate activities in global agrifood governance. The contributing analyses have delineated the role of corporate power in the Codex Alimentarius, the establishment of organic farming standards, international food aid, the way retailers have started to govern supply chains, international rules on transparency in the food chain, and various aspects of the regulation of genetically modified organisms (GMOs). Second, we have aimed to analyze the implications of this expansion of corporate influence in agrifood governance for sustainability. Here, the contributing authors have highlighted potential improvements in some aspects of sustainability, but also detriments, such as threats to rural livelihoods especially in developing countries, as well as to biodiversity and food safety. Moreover, they have assessed the differing discourses on sustainability used by corporate actors in the major political and societal debates concerning food and food regulation. Below, we reflect on some of the common themes that emerge from the analyses presented in this book.

Corporate Power and Its Influence on Global Agrifood Governance

In the effort to delineate the various ways corporations exercise power in global agrifood governance, the case studies have drawn on a common, multidimensional framework of power laid out in chapter 1. Specifically, they have differentiated between the instrumental, structural, and discursive forms of power employed by corporations. The role of instrumental power, for example, becomes very clear in Clapp's analysis of the lobbying by the grain and shipping industries of the US Congress and in particular the participation of industry representatives in hearings and committees. Likewise, the role of structural power is pinpointed in the analysis by Fuchs, Kalfagianni, and Arentsen of the power retail corporations can exercise via private standards due to their extensive reach. Similarly, the role of discursive power is revealed, for example, in Smythe's and Sell's analyses of corporate support for "scientific knowledge" and "secure property rights" frames over "uncertainty" or "right-to-know" frames. Each case study has delineated the presence and exercise of at least two forms of corporate political power, thereby revealing the complex strategies employed by corporations in pursuit of their goals in global agrifood governance (see table 10.1).[1]

With the common power-theoretic framework outlined in chapter 1, the book has been able to focus on an aspect not sufficiently analyzed in the existing literature on the global food system. Specifically, the case studies demonstrate that corporations have not only been the driving force behind the globalization of the food system, but are also key actors in its governance in various ways. Moreover, the analyses reveal the complex and multifaceted political (not just economic) strategies employed by corporate actors in pursuit of their interests in the governance of the global food system, and the types of power and channels of its exercise that provide the opportunities for corporate influence in global agrifood governance.

Table 10.1
The case studies' focus and findings

	Dimensions of power	Aspects of sustainability	Link	Structural context	Other actors	Policy implications
Fuchs et al.	Structural, discursive	Health, environment, social; discourses: food safety	Improvements in food safety; reduction in food security/ social sustainability	Oligopoly; Expansion in private governance institutions	Consumers (North), NGOs	Need to include social sustainability in standards; necessity of public frame for private governance and of regulation of retail power
Scott et al.	Structural, discursive	Health, environment, social	Loss of local knowledge and practices relating to organic production	Oligopoly; Facilitating national regulation	Local farmers, NGOs, state actors	Create opportunity for wider participation in standard setting process
Smythe	Instrumental, discursive, (structural)	Safety, environment, right to know; discourses: scientific knowledge, uncertainty	No agreement on labelling of GMOs, transfer of risk to consumers and producers	Oligopoly; Global food standards and trade regimes	States, NGOs	Need for improvements in transparency and participation via transparent and accountable regulatory agencies
Clapp	Instrumental, structural, discursive	Food security and safety, environment, economic/ social, discourses: AIDs, corruption, market access	Food security improved only in crisis situations	Oligopoly; global food aid and trade regimes; national regulations in US	State actors, NGOs	Provision of cash-based aid instead of in-kind aid

Table 10.1 (continued)

	Dimensions of power	Aspects of sustainability	Link	Structural context	Other actors	Policy implications
Williams	Discursive, (instrumental, structural)	Food safety and security, environment; discourses: food security, efficiency, food quality	Myth of improved food security creates favorable policy environment; risk assessment and distribution	Oligopoly; Global trade regime; facilitating national regulations	State actors, universities, consumers/ NGOs	Recognize alternative discourses in policy debates on GMOs
Sell	Structural, instrumental, discursive	Environmental, social, economic; discourses: secure property rights	Loss in countries' autonomy in sustainability related choices; loss in access to knowledge	Oligopoly; Global intellectual property rights regime/different fora	State actors, farmers, universities, NGOs	Need larger role of universities, and public funding of research; accessibility of technology
Falkner	Structural, discursive, instrumental	Various GMO related discourses, e.g., crop variety	Competition in sustainability frames	Oligopoly; multilateral regimes (e.g., Cartagena Protocol); national/ regional regulation	Consumers/ NGOs, farmers, EU, AG traders	Business conflict creates space for public actors and civil society
Newell	Discursive, institutional, material	Environmental, social, economic	Rising chemical imports, deforestation, land concentration, decreasing employment	Oligopoly; global and regional trade and property rights regimes	Farmers, state actors	Create spaces to debate openly the complete costs and benefits of existing and alternative agricultural development trajectories

As pointed out in the introductory chapter, the types of power constituting our three-dimensional framework are sometimes easier to separate analytically than empirically. Yet the analytic differentiation is important, because it allows us to reveal the various means and points of political intervention as well as the sources of influence of corporations in global agrifood governance. Williams's chapter, for example, while focusing on the discursive element of corporate power regarding questions of food security and GMOs, also illustrates the overlap and reinforcement of discursive power with the structural and instrumental dimensions. At the same time, we do not claim that our power-theoretic framework explains everything. The individual case studies have brought in additional theoretical approaches as necessary for an understanding of specific cases.

The chapters together present as comprehensive, and yet as differentiated, a picture as possible of corporate power in the design, implementation, and evaluation of various rules and regulations in the global food system. The case studies focus on different types of corporate actors, different issues in, and locations of, global food governance, as well as different forms of governance. A number of authors focus on the role of global biotechnology firms, while others explore the US grain and shipping industry, retail corporations, or agricultural traders. Likewise, the case studies analyze global developments and their implications for local developments, national politics with implications for global governance, or regional developments and their impact on national or global aspects of agrifood governance. The chapters by Newell and Clapp, for example, demonstrate the importance of national-level politics of a particular agrifood governance issue that has global implications, while Scott, Vandergeest, and Young analyze the interaction of global developments with national and local settings. The case studies also highlight the growing role of corporations in public governance and private governance mechanisms, as well as the interaction between public and private governance. The chapters that examine private governance, including those by Scott, Vandergeest, and Young and by Fuchs, Kalfagianni, and Arentsen, also show how it can supplement or either officially or unofficially bypass national and global public governance.

Implications for Sustainability

The various case studies explore different facets of sustainability as well as sustainability discourses in relation to corporate power and its influence on global agrifood governance. As table 10.1 shows, the definition of sustainability employed in our inquiry is rather broad, including not only the environmental aspects, but also social, economic, and political dimensions. This broad view of sustainability enabled the case-study analyses to provide as complete a picture as possible of the wider implications of the exercise of, and attempted exercise of, corporate power in global agrifood governance.

The chapters identify the impact of corporate influence on environmental quality, crop variety, farmer livelihoods, as well as on the distribution of risk and the value of local practice and knowledge. For example, food aid policy determined by corporate power in a way that promotes in-kind aid can result in harm to farmer livelihoods in recipient countries. Similarly, corporate power has prevented an agreement on labeling GMOs, thus neglecting consumers' right to know as well as their opportunities for choice. Likewise, corporate retail power negatively affects farmer livelihoods as well as local knowledge and practice with respect to organic production. This is a particularly striking result, because it indicates a negative effect on sustainability even of measures officially aimed at improving it.

In addition to the broader impacts on sustainability, the various case studies demonstrate that corporate actors also attempt to define sustainability in their interest in political debates in the public arena. For example, corporate actors have promoted discourse emphasizing scientific knowledge rather than uncertainty in the case of GMOs, as shown in the chapters by Smythe and Williams. As shown by Fuchs, Kalfagianni, and Arentsen as well as by Scott, Vandergeest, and Young, corporate actors have also attempted to narrow the concept of sustainability to mean food safety, enabling retailers to market certain products as "green" or "safe", but entailing potentially negative social or economic impacts for producers. Corporate actors have also attempted to use sustainability-related discourse such as "food security" to promote their products, even though the ability of the relevant activities (such as the promotion of GMOs, as discussed by Williams) to provide safe, nutritious, and affordable access to food for the world's poor is far from clear.

Importance of the Structural Context

The case studies share their identification of certain structural contexts in which corporate power in global agrifood governance is embedded, and which, we would argue, actually provide a precondition for the extent of this power. These structural contexts include both the market setting and the ideological setting. Corporate political activity in all of the case studies takes place in an oligopolistic (or oligopsonistic) setting. This is perhaps not surprising, because the globalization of the food system has been paralleled by a trend toward capital concentration. It is nevertheless worth mentioning this fact, because it shows that these oligopolies in the various sectors of the food system provide a basis not just for economic power, but also for political power, or put differently, for an absence of a plurality in the political contest. In turn, the finding strongly suggests the need for public intervention in the case of monopolistic or oligopolistic market conditions not just for economic, but also for political reasons.

The case studies also delineate the ways corporate power is embedded in and at a minimum facilitated by neoliberal political economic systems. For example, the global trading, intellectual property rights, biosafety, and food aid regimes all provide the space for corporations to become powerful political actors by prioritizing economic and investor interests and cloaking them in legitimacy. The trumping of the public interest by private interests, then, is partly predetermined by this setting, in which economic objectives tend to be considered more than social or environmental ones.

The Role of Other Actors

The case studies have revealed the frequently pivotal interaction between corporate power and state power in both developed and developing countries. We had acknowledged in the introduction that one should not think of corporate power as existing in a vacuum—that is, that corporate influence often will depend on state-business relationships. Yet, the extent to which public actors create rules and regulations serving corporate interests in food governance is astonishing. State actors pursue corporate interests not only at the national level, but also in multilateral forums. Public regulation can facilitate the successful exercise of corporate power in some cases. But the absence of public regulation can also help corporations insofar as a regulatory vacuum leaves room for private governance schemes.

Following on this point, many of the case studies actually provide evidence against arguments that public actors are powerless vis-à-vis corporate actors. Rather, they underline the complicit role of public actors and suggest the need to inquire into its causes. Scott, Vandergeest, and Young show this role of the state in the corporatization of the organic sector in Southeast Asia, for instance, while Newell documents it with respect to GMOs in Argentina. Such an inquiry into the strengthening of opportunities for corporate influence, in turn, leads us back to the impact of neoliberal norms as well as the successful exercise of corporate power with respect to national governments and bureaucracies.

Besides looking at the part played by public actors, many of the case studies have delineated the crucial role of NGOs. Here, however, we notice some differences between these roles. In food aid policy, for instance, some of the NGOs have long supported the same policies as the relevant corporate actors and only recently assumed a more critical position, due to pressure exerted on them by their membership and other NGOs. In other cases, NGOs have been important contesters of corporate power, even if frequently on the losing end (though not always, as Falkner shows), due to the asymmetries in resources between them and the large corporations.

Additional Insights

Besides offering these common insights, the various case studies have contributed interesting individual findings that add to our understanding of the role of corporate power in global agrifood governance. Sell, for instance, has demonstrated that forum shifting is a crucial strategy employed by corporations in their pursuit of GMO-friendly international regulation. Clapp has shown that the institutional setup of food aid policy in the United States, which involves congressional approval each time there is a change in policy, is particularly susceptible to penetration by corporate power (and other lobbyists). Likewise, Scott, Vandergeest, and Young as well as Fuchs, Kalfagianni, and Arentsen have drawn our attention to shifts in power among corporate actors in global agrifood governance, away from producers to retailers, and to the expansion in private governance. Newell's analysis, in turn, reveals how the politics of poverty interact with corporate power in GMO-related agrifood governance in Argentina.

In addition, the analyses have identified likely determinants of the success of the exercise of corporate power. Falkner, in particular, suggests that corporate power is constrained in cases of conflicting interests within the business community. If the business community does not speak with a unified voice, political space for civil society and public actors emerges. The extent to which this is a source of hope for observers critical of corporate influence on agrifood governance is debatable. On the one hand, Falkner's case study demonstrates the failure of (some) corporations to achieve their political goals. On the other hand, if the only remaining enemy of business influence is business itself, as Berry (1997) has argued as well, the strategies and likelihood of success for civil society in cases in which such business conflict is absent and cannot be created clearly are limited. An additional determinant of the successful exercise of corporate power is identified by Williams, who argues that the influence of corporate discursive power in agrifood governance is particularly great in cases of nonconsensual science.

Areas for Further Research

A number of issues requiring further research arise from this study. First, a more systematic understanding of corporate choices in terms of the political strategy used, in particular the intervention points targeted, would be valuable. The case studies suggest some preliminary answers here, but cannot provide a complete picture yet. In terms of the type of power exercised, it is astonishing that almost all of the case studies delineate the simultaneous exercise of instrumental, structural, and discursive power as well as the interaction between them.[2] Thus, they highlight the need for future research on corporate power in food governance to take these various dimensions and their interaction into account. In addition, it seems likely that the exercise of corporate power in other policy fields is similarly multidimensional, and so the need to consider all three types of power would apply there as well.

In addition, the discursive dimension of corporate power and, specifically, the link between power and knowledge, needs further exploration. Discursive power is particularly potent because its targets can range from details of specific policy designs to actor interests and identities to broader societal norms. Moreover, in the media and information age, resource asymmetries between corporate actors and others add to the relative discursive power corporations can exercise. Because of the diffuse nature and low visibility of discursive power, it can be particularly difficult to contest for those with fewer resources.

The issue of the democratic legitimacy of global food governance also requires further scholarly attention. Given the analyses' findings on the powerful influence of corporate interests on food politics and policy and its often-negative implications for the sustainability of the global food system, the question of how participation, transparency, and accountability can be improved needs to be investigated to provide a foundation for appropriate political responses. This applies to private food governance in particular. Some of the literature on private governance has argued that such governance schemes acquire political legitimacy through the expertise provided by the private actors involved and their effectiveness in providing the intended output. The implications of the analysis by Fuchs, Kalfagianni, and Arentsen, however, are that this notion of "output legitimacy" is seriously flawed. An evaluation of the effectiveness of private governance institutions in achieving certain goals depends on the definition of the goal applied (in this case sustainability). At the same time, such an evaluation cannot measure the result simply against the goals set by the initiative itself, as frequently suggested by proponents of private governance institutions and the concept of "output legitimacy." After all, these goals may be set very low or might even be illegitimate themselves from a normative point of view. Questions about the democratic legitimacy of private governance thus need to be raised again and one can only caution against too naively attributing "output legitimacy" to such institutions.

Policy Implications

What, then, are the implications of the case studies' findings for policy and politics? Given the at best ambivalent implications of the exercise of corporate power in global agrifood governance, we have to ask ourselves what strategies for limiting the negative effects of corporate power could look like. The various case studies suggest answers to this question as well. Some of them point out issue-specific sustainability measures. Thus, Sell argues for the need for public funding for agricultural research, and Clapp for the provision of cash-based food aid instead of in-kind food aid. Likewise, Fuchs, Kalfagianni, and Arentsen demonstrate the need to include social criteria in private retail governance and highlight the need for a corresponding public framework for private governance schemes. Newell, in turn, emphasizes the need to create spaces to openly debate the environmental and social costs and benefits of existing and alternative agricultural development trajectories.

At the same time, some of the chapters suggest more general measures to limit corporate power in global agrifood governance. Smythe, for example, argues for improvements in transparency and participation via transparent and accountable regulatory agencies. In addition, the insight drawn from the structural preconditions of corporate power in agrifood governance discussed above should be mentioned here again: the need to prevent the existence of monopoly or oligopoly power by corporate actors to allow for both a plurality of voices in politics (in fact, to allow for Falkner's business conflict) and to reduce asymmetries in power among the different actors involved.

A major policy implication is the need to improve the democratic legitimacy of global agrifood governance. Relevant measures in this context have to address questions of participation, transparency, and accountability. Supportive measures are needed to ensure the sufficient participation of not-for-profit interests in the political process, because civil society faces substantial resource constraints as well as collective action problems. For improvements in transparency, measures allowing the provision and easy accessibility of information on political processes and the contents of political decisions are needed, all the more given that we are living in an era of information overflow in which information is often provided to draw attention away from other information. The accountability of public and private actors also needs to be ensured. Participation and transparency are preconditions for this accountability. But additional measures are required here as well.

One should not assume that any of these measures will be achieved easily. If public actors have failed to adopt the appropriate food-related policies in the past or have not provided more stringent regulation of corporate political power, why should we expect this situation to change? After all, the assumption that public governance can refrain from succumbing to too much corporate influence or that it can provide a sound framework for private governance presupposes a well-functioning democratic system. Thus, our argument for public governance should not be understood as a claim that that will be easily achieved or as naively positing that any public governance is better than private governance. In fact, we should probably expect public actors to take action in this context only if significant pressure is exercised by civil society. Of course, civil society faces the challenge of limited resources (human, financial, organizational). Nevertheless, concerted action focusing on this very issue may provide a more promising strategy than the current overextension and exhaustion of a number of underfunded NGOs across a broad range of food-related forums and negotiations.

Acknowledgments

The authors are grateful for research assistance provided by Frederike Boll, Stephan Engelkamp, Katharina Glaab, and Richard Meyer-Eppler.

Notes

1. Newell has emphasized sources rather than types of power.

2. While neither Fuchs, Kalfagianni, and Arentsen nor Scott, Vandergeest, and Young have chosen to focus on instrumental power due to the more visible developments in the structural and discursive power of retail corporations, instrumental power is exercised by these actors as well. This is demonstrated in the country case studies provided by Scott and colleagues, for instance.

Reference

Berry, Jeffrey. 1997. *The Interest Group Society*. New York: Longman.

Index